I0489849

NUREG/CR–5560
RV

Aging of Nuclear Plant Resistance Temperature Detectors

Manuscript Completed: March 1990
Date Published: June 1990

Prepared by
H. M. Hashemian, D. D. Beverly, D. W. Mitchell, K. M. Petersen

Analysis and Measurement Services Corporation
AMS 9111 Cross Park Drive
Knoxville, TN 37923–4599

Prepared for
Division of Engineering
Office of Nuclear Regulatory Research
U.S. Nuclear Regulatory Commission
Washington, DC 20555
NRC FIN D2039

ISBN-13: 978-1499558982; ISBN-10: 1499558988

This manuscript has been authored by a contractor of the U.S. Government under Grant No. D-2039. Accordingly, the U.S. Government has a nonexclusive, royalty-free license to publish or reproduce the published form of this contribution, or allow others to do so, for U.S. Government purposes.

ABSTRACT

A comprehensive research and development project was performed over a 30 month period to quantify the effects of normal aging on performance of nuclear safety-related RTDs. The work involved laboratory testing of 72 nuclear grade RTD elements representing several from each of the four U.S. manufacturers. The limit for the initial accuracy of these RTDs was established and a procedure for performing precise calibration was developed. Experimental aging of 30 of these RTDs in simulated reactor conditions resulted in five failures and six major calibration shifts. Two failures occurred in thermal aging, one in vibration aging, one in humidity aging, and one in thermal cycling. The remaining 19 RTDs performed well during the aging tests, maintaining a drift band of ± 0.2°C.

The shelf-life drift of RTDs was also quantified. This involved testing 24 RTDs that have been in normal storage in various nuclear power plants for periods of one to five years and 21 RTDs that were aged in the project for storage effects. The test results for these 45 RTDs have shown a shelf life drift band of ± 0.1°C. Most of the storage drifts, the failures, and the normal aging drifts were found to occur in the first few months of aging. A potential remedy is to burn-in the RTDs before they are calibrated and installed in the plant.

The performance of nuclear plant RTDs is evaluated by response time testing in addition to calibration. Response time and calibration of RTDs are independent and are therefore tested separately. The nuclear industry has about ten years of experience with RTD response time resulting from periodic in-situ measurements made in about 60 PWRs at least once every fuel cycle. Representative results of these measurements were reviewed to identify the range of achievable response times and the response time degradation modes.

The project did not reveal any unanticipated or major systematic aging problem in the performance of the RTDs tested. The nuclear industry's practice for verifying adequate RTD accuracy and response time is to perform on-line cross calibration and Loop Current Step Response tests at least once every fuel cycle. In light of the data obtained throughout this study, this approach is reasonable for management of aging of RTDs which do not have any major design, fabrication, or installation deficiencies. RTDs that consistently maintain a suitable calibration and response time, as determined by periodic testing, can be used in the plant for their qualified life as specified by the manufacturer. The manufacturer's specification for qualified life of nuclear grade RTDs typically range from 10 to 40 years depending on the manufacturer and the conditions in which the RTDs are used.

ABSTRACT

TABLE OF CONTENTS

APPENDICES

LIST OF FIGURES

LIST OF TABLES

1. INTRODUCTION

This report presents the details of a comprehensive research and development project on aging of narrow range RTDs used in the primary coolant system of pressurized water reactors (PWRs). The goal was to establish the long term performance limits of these RTDs in order to verify that objective and adequate measures are implemented to ensure safety.

The project was conducted in two phases. Phase I, a six month feasibility study, was completed in June 1987. The results are published in the NUREG/CR-4928 report entitled, "Degradation of Nuclear Plant Temperature Sensors"[1]. This work demonstrated the need for a long-term Phase II project which was conducted over a 30-month period beginning in October 1987. The results of this Phase II work are reported herein.

The first step in the Phase II project was to set up a laboratory with calibration and aging equipment and to obtain nuclear grade RTDs. Twelve new RTDs were obtained, three from each of the four most commonly used manufacturers. Together with eight RTDs from Phase I and 31 RTDs provided by interested utilities, the project was started with 51 nuclear grade and 17 commercial grade RTDs. The commercial grade RTDs were included for comparison purposes. Of the 51 nuclear grade RTDs, 21 were dual element providing a total of 72 independent RTD elements. Of these, 30 elements were used in one or more of the five aging categories shown in Table 1.1. The remaining 42 elements were used in various other tests performed in the project.

Next, a computer-based automatic calibration and monitoring system and procedure were developed. The RTDs were calibrated and placed in two furnaces at approximately 320°C, the primary coolant temperature in most PWRs. The RTDs were monitored in the furnaces using a computer scanning system which measured and stored their loop resistance, insulation resistance, open circuit voltage, and lead wire resistance. These measurements helped identify and characterize the failures when they occurred. Once every one or two months, the RTDs were removed from the furnaces and calibrated to quantify any drift. The thermal aging process was continued for 18 months, equivalent to a typical PWR fuel cycle. Of 30 RTD elements tested for thermal aging, two failed, six showed drift in the range of 0.6 to 3.0°C, but the remaining 22 drifted less than 0.2°C over the entire thermal aging period.

The RTDs were then stored at room temperature, pressure, and humidity, and periodically calibrated to identify shelf-life drift. This work was performed over a four month period. The results showed that the RTDs are not immune to degradation during storage. This problem can be resolved by recalibrating the RTDs shortly before they are installed in the plant.

TABLE 1.1

Number of Nuclear Grade RTDs Included in Each Aging Category

RTD	Thermal Aging	Vibration Aging	Humidity Aging	High Temperature	Thermal Cycling
1	◊	◊		◊	◊
2	◊				
3	◊				
4	◊				
5	◊				
6	◊		◊	◊	◊
7	◊				◊
8	◊	◊		◊	◊
9	◊	◊		◊	◊
10	◊				◊
11	◊		◊		
12	◊		◊	◊	◊
13	◊		◊	◊	◊
14	◊	◊		◊	◊
15	◊	◊		◊	◊
16	◊		◊		
17	◊	◊		◊	◊
18	◊	◊		◊	◊
19	◊	◊		◊	◊
20	◊	◊		◊	◊
21	◊		◊	◊	◊
22	◊		◊	◊	◊
23	◊				
24	◊		◊	◊	◊
25	◊				
26	◊	◊			
27	◊		◊	◊	◊
28	◊		◊	◊	◊
29	◊		◊		
30	◊				
Number Tested	30	10	11	17	19

The aging of the RTDs was continued to identify the effects of vibration, humidity, mechanical shock, high temperature, and thermal cycling. These tests resulted in three more failures, but did not increase the average drift of the RTDs beyond that of thermal aging. Several commercial grade RTDs were also aged and tested for comparison with nuclear grade RTDs. The results showed that the average response time and calibration stability of nuclear grade RTDs is about twice as good as that of the commercial grade RTDs.

The project addressed the following additional topics: sources of errors in RTD calibration, factors affecting RTD accuracy and response time, failures of RTDs as reported in the LER and NPRDS databases, and the International Temperature Scale of 1990 and its impact on temperature measurements in nuclear power plants.

2. TERMINOLOGY

The common terms used in describing the physical characteristics and performance of nuclear plant RTDs are defined in this section. Some of these terms are further explained in the body of the report.

- Accuracy. The maximum positive or negative difference that may exist between the actual process temperature and the temperature indicated by the RTD. It is usually measured as inaccuracy and expressed as accuracy. Also called uncertainty, error, or calibration accuracy. This includes calibration errors as well as inherent RTD errors such as hysteresis, repeatability, and self heating.

- Calibration. The relationship between RTD resistance and temperature. A chart which lists the resistance of an RTD as a function of temperature is called a calibration chart or calibration table. A plot of resistance versus temperature is called a calibration curve. For the levels of accuracy required in the nuclear industry, the calibrations must be uniquely determined for each RTD.

- Commercial Grade RTD. A general purpose RTD made for general industrial applications as opposed to nuclear safety-related applications.

- Cross Calibration. Comparison of the average indication of a group of RTDs with each individual indication to check for consistency and identify the outliers. Cross calibration is a method for on-line testing of accuracy of installed RTDs.

- Degradation. Gradual changes in calibration or response time of an RTD. Response time changes are usually called degradation and calibration changes are called drift or shift.

- Drift. Changes in accuracy over time. Also called calibration drift, drift rate, shift, calibration shift, stability, or instability.

- Error. Synonymous with uncertainty, inaccuracy, or accuracy.

- Failure. An RTD is said to have failed if its sensing element or any of its extension leads have opened or shorted to the sheath, or in the case of this report, its calibration at any point within the 0 to 300°C range is shifted by more than 5°C.

- Insulation Resistance. The electrical resistance between the RTD sensing element or any extension lead and the sheath.

- LER (Licensee Event Report) Database. A compilation of reportable failures of certain components in nuclear power plants.

- <u>NIST (National Institute of Standards and Technology)</u>. Formerly known as National Bureau of Standards or NBS. The calibration of RTDs and associated test and measurement equipment are traced to NIST using transfer standards such as SPRTs and resistance and voltage standards.

- <u>Normal Aging</u>. Degradation of RTD performance over time while subjected to normal environments and operating conditions.

- <u>NPAR (Nuclear Plant Aging Research Program)</u>. A program initiated by the NRC in the early 1980's to provide an understanding of aging of components, systems, or structures in nuclear power plants.

- <u>NPRDS (Nuclear Plant Reliability Data System)</u>. A compilation of reports of failure of certain nuclear power plant components which nuclear utilities voluntarily file with the Institute of Nuclear Power Operations (INPO).

- <u>Nuclear Grade RTD</u>. A platinum RTD designed for use in safety related applications in nuclear power plants.

- <u>On-line Testing</u>. Remote testing of installed RTDs while the plant is operating. Also called in-situ testing.

- <u>Performance</u>. A general term used to refer to the static (calibration or accuracy) and dynamic (response time) performance of an RTD.

- <u>R vs. T Curve</u>. Resistance versus temperature relationship, curve, table, or chart of an RTD.

- <u>Random Error</u>. Errors whose value can be positive or negative with respect to the actual temperature. Random errors are sometimes called uncertainty.

- <u>Repeatability</u>. The ability to obtain the same calibration with the same RTD. Repeatability is the maximum difference, throughout the RTD operating range or at a given temperature, between the results of repeated calibrations of the same RTD using the same equipment and procedure. Also called precision.

- <u>Response Time</u>. The time required for an RTD to reach 63.2 percent of its final value following a step change in temperature. Also called time constant even though time constant is meaningful only for a first order system and RTDs are not necessarily first order.

- <u>RTD (Resistance Temperature Detector)</u>. A term used in referring to industrial resistance thermometers. If the sensing element of the RTD is made of platinum wire, the RTD is called platinum resistance thermometer, PRT, or platinum RTD.

- <u>Self Heating</u>. The phenomenon in which heat is generated in an RTD due to the electric current used for measurement of its resistance.

- <u>Sensing Element</u>. The wire inside the RTD whose resistance changes with temperature.

- <u>Shift</u>. Changes in the resistance versus temperature relationship of an RTD, also called drift. Shift implies a total change with time at the end of a period or a test while drift implies gradual changes.

- <u>SPRT (Standard Platinum Resistance Thermometer)</u>. Also called standard RTD. SPRTs are calibrated at the National Institute of Standards and Technology (NIST) and used as a transfer standard for laboratory calibration of industrial RTDs.

- <u>Stability</u>. The ability of the RTD to maintain its accuracy. Stability is quantified by drift or drift rate (°C/year).

- <u>Systematic Error</u>. Additive errors. A constant error or bias.

- <u>Thermowell</u>. A protective jacket (tube) used to protect the RTD from the process fluid and to permit easy replacement.

- <u>Well-Type RTDs</u>. RTDs that are designed to be installed in a thermowell.

- <u>Wet-Type RTDs</u>. RTDs that are installed directly into the process fluid as opposed to being installed in a thermowell. Also called direct immersion RTDs.

3. BACKGROUND

The interest in performance testing of nuclear plant RTDs began when Regulatory Guides 1.118 and 1.105 were issued by the NRC in the mid 1970's recommending periodic sensor response time testing and calibration. In response to these regulatory guides, the Electric Power Research Institute (EPRI) funded two projects to develop in-situ methods for response time testing and calibration of RTDs.[2,3] These projects resulted in development of the Loop Current Step Response (LCSR) method for response time testing and the Johnson noise method for calibration. The LCSR method has been successfully implemented in many plants, but the Johnson noise method has encountered difficulty and is therefore not currently used.

There have been some discussions as to why periodic calibration and response time testing is necessary for RTDs that have been qualified by their manufacturers for up to 40 years of service. Although the environmental and seismic qualification tests are useful in determining accident survivability, experience has shown that these tests do not generally provide reliable information about long-term performance of RTDs under normal use. The experience available with response time testing of nuclear plant RTDs has shown that major response time increases can occur with normal use. With respect to calibration drift however, adequate laboratory research or in-plant experience is not yet available. The work reported herein is the first systematic research on stability of RTDs in the temperature range of 0 to 400°C which is the range of interest in the primary coolant system of PWRs. Other pertinent research is that of the NIST which concentrated on the performance of SPRTs[4], and a few published results which involve temperature ranges other than 0 to 400°C.[5,6]

4. PROJECT OBJECTIVES

The goal of this project was to establish the long-term performance limits of RTDs. The project answered the following specific questions.

- How well can an RTD be calibrated?

- How long does an RTD maintain its calibration?

- How often should RTDs be recalibrated or replaced?

- What is an acceptable procedure for laboratory calibration of RTDs?

- Is there drift during storage?

- What are the acceptable methods for in-situ calibration and in-situ response time testing of RTDs?

- What factors affect the RTD response time and what is the range of response times that can be achieved and maintained with RTDs?

The answers to these questions help determine the performance limits of RTDs, frequency of testing, and acceptable test methods.

5. DESCRIPTION OF NUCLEAR PLANT RTDs

Two groups of RTDs are used in nuclear power plants: direct immersion (or wet-type) and thermowell-mounted (or well-type). These installations are illustrated in Figures 5.1 and 5.2. The advantage of direct immersion RTDs is a better response time and the disadvantage is difficulty in replacement. The advantage of well-type RTDs is ease of replacement and the disadvantage is a larger response time than direct immersion RTDs and succeptability to response time degradation due to changes in the RTD/thermowell interface.

There are five manufacturers of safety-related RTDs for nuclear power plants. These are Conax Corporation, RdF Corporation, Rosemount Inc., Sostman Company, and Weed Instrument Company. Examples of RTDs supplied by these manufacturers and plants where they are currently used are given in Table 5.1. A photograph of a typical well-type RTD and its thermowell is shown in Figure 5.3. A photograph of three wet-type RTDs is shown in Figure 5.4. Typical characteristics of nuclear grade RTDs are summarized in Table 5.2. Sostman RTDs were not included in this project because they are currently used in only a few plants.

The number of RTDs used in a plant varies widely even among identical plants. The range is about twelve to thirty RTDs depending on the number of primary coolant loops and the installed spares. Additional RTDs are sometimes used to compensate for temperature stratification problems. Both single element and dual element RTDs are used. Dual element RTDs have two independent sensing elements in the same sheath. One element may be used for control and the other for safety-system applications. In some plants, the second element is used as an installed spare.

Two designs are used by manufacturers for achieving fast response with well-type RTDs. In one design, the sensing tip of the RTD and thermowell is tapered (Figure 5.5). In another design, the sensing tip of the RTD is flat, but silver brazed, silver plugged, or silver plated for improved response time (Figure 5.6). Silver is soft and acts to fill the gap at the RTD/thermowell interface and thereby results in a faster response. Gold is also used for plating of RTDs but silver is more common.

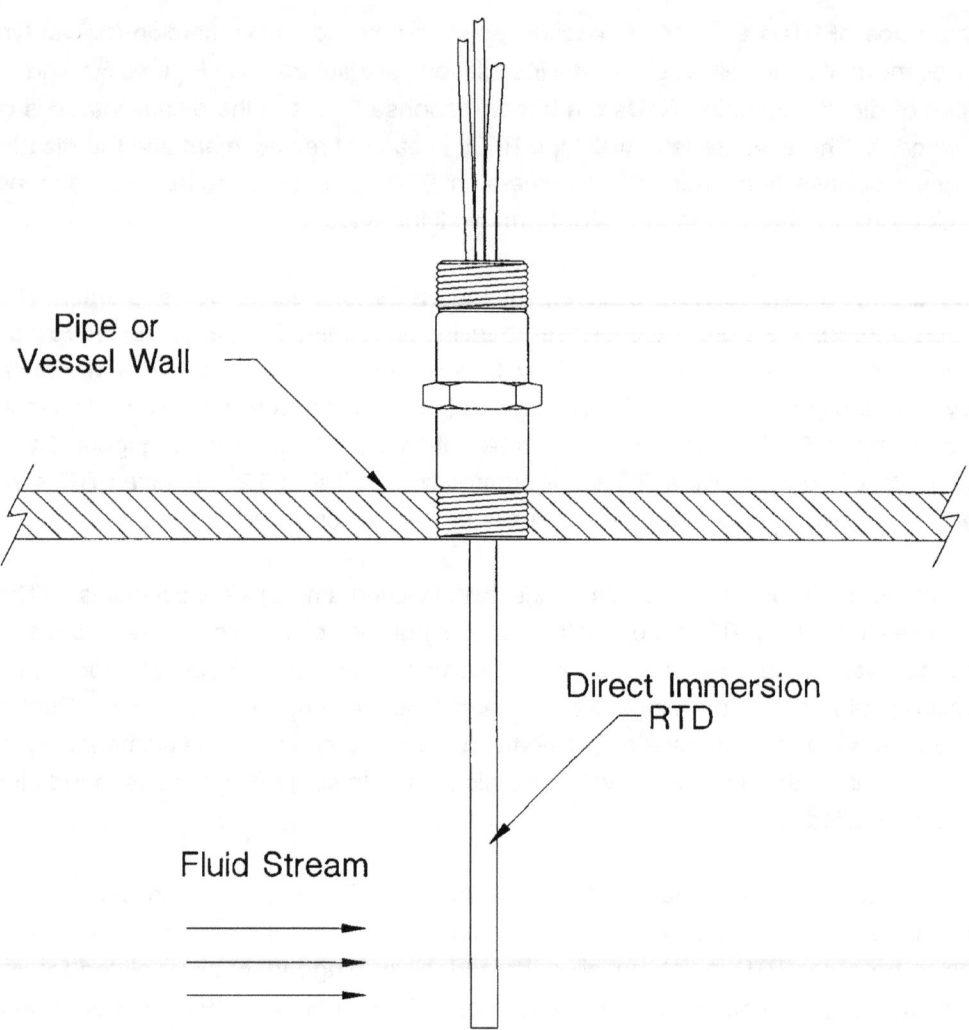

Figure 5-1. Installation of Direct Immersion RTDs.

- **10** -

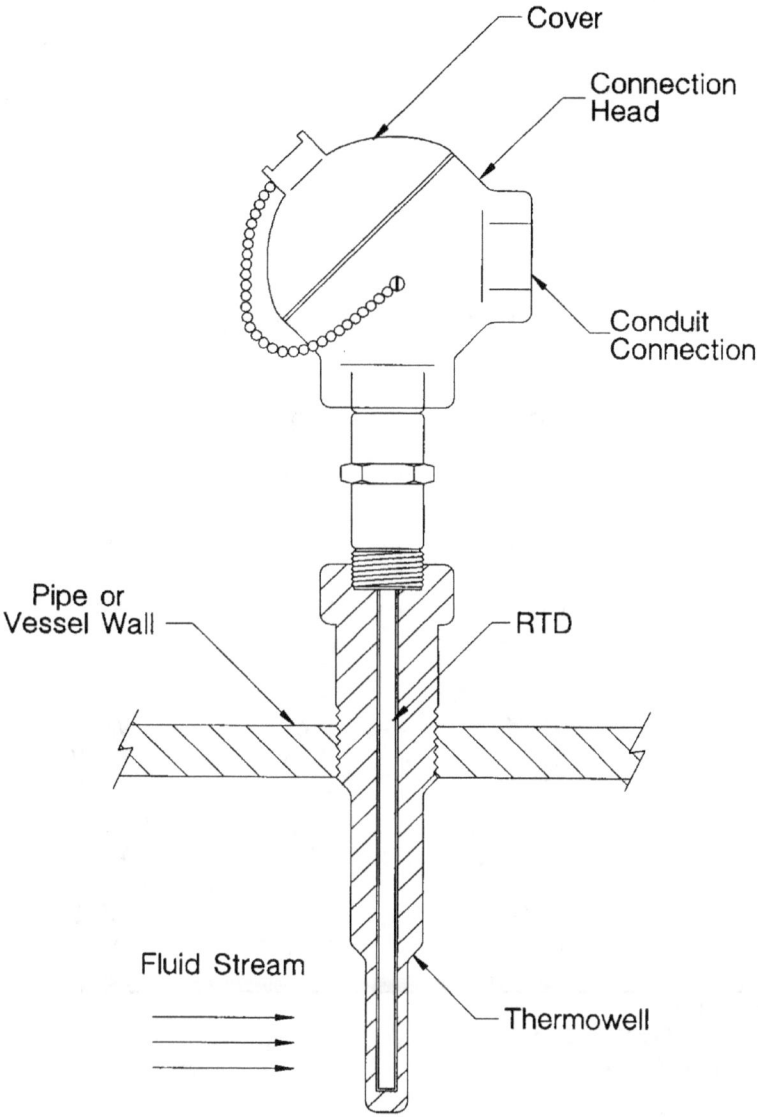

Cover

Connection
Head

Conduit
Connection

Pipe or
Vessel Wall

RTD

Fluid Stream

Thermowell

Figure 5-2. Installation of Well-Type RTDs.

TABLE 5.1

Examples of RTDs Used in Nuclear Power Plants

Manufacturer	Model	Type	Plant
Conax	7N10	Well	CE
RdF	21204	Wet	W
	21232	Well	W
	21458	Well	CE
	21459	Well	CE
Rosemount	104	Well	CE
	176KF	Wet	W
	177HW	Well	B&W
	177GY	Wet	B&W
Sostman	11834	Wet	W
Weed	N9004	Well	CE & W
	N9007	Wet	W

W   :   Westinghouse
CE   :   Combustion Engineering
B&W :   Babcock & Wilcox

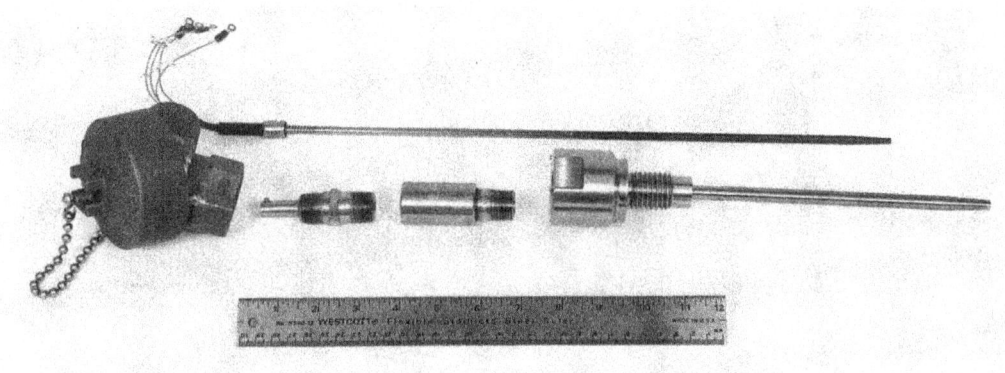

Figure 5-3. Photograph of a Typical Well-Type RTD and its Thermowell.

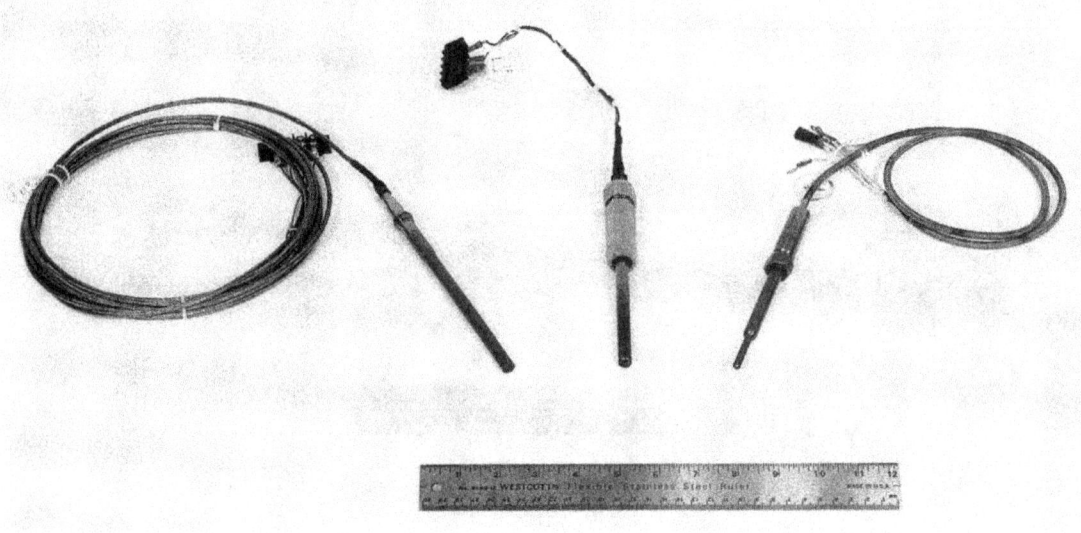

Figure 5-4. Photograph of Three Wet-Type RTDs.

TABLE 5.2

Typical Characteristics of Nuclear Plant RTDs

Average Length	30 - 60 cm Well-type 12 - 18 cm Wet -type
Average Diameter	0.6 - 1.0 cm RTD 1.0 - 2.0 cm Thermowell
Average Weight	100 to 250 grams RTD 300 to 3000 grams Thermowell
Sheath Material	Stainless Steel or Inconel
Sensing Element	Fully Annealed Platinum Wire
Ice Point Resistance(R_0)	100 or 200 Ohm
Temperature Coefficient(α)	0.003850 $\Omega/\Omega/°C$ Regular grade 0.003902 $\Omega/\Omega/°C$ Premium grade
R vs. T Curvature (δ)	1.5 (°C)
Temperature Range	0 to 400°C
Insulation Resistance(IR)	Greater than 100 megohm at room temperature measured with 100 VDC
Response Time (1 m/sec water)	0.5 - 5 sec Wet -type 4 to 8 sec Well-type
Self Heating Index (1 m/sec water)	2 to 10 Ω/W

---

cm = *centimeter*
Ω = *ohm*
W = *watt*
m/sec = *meter per second*

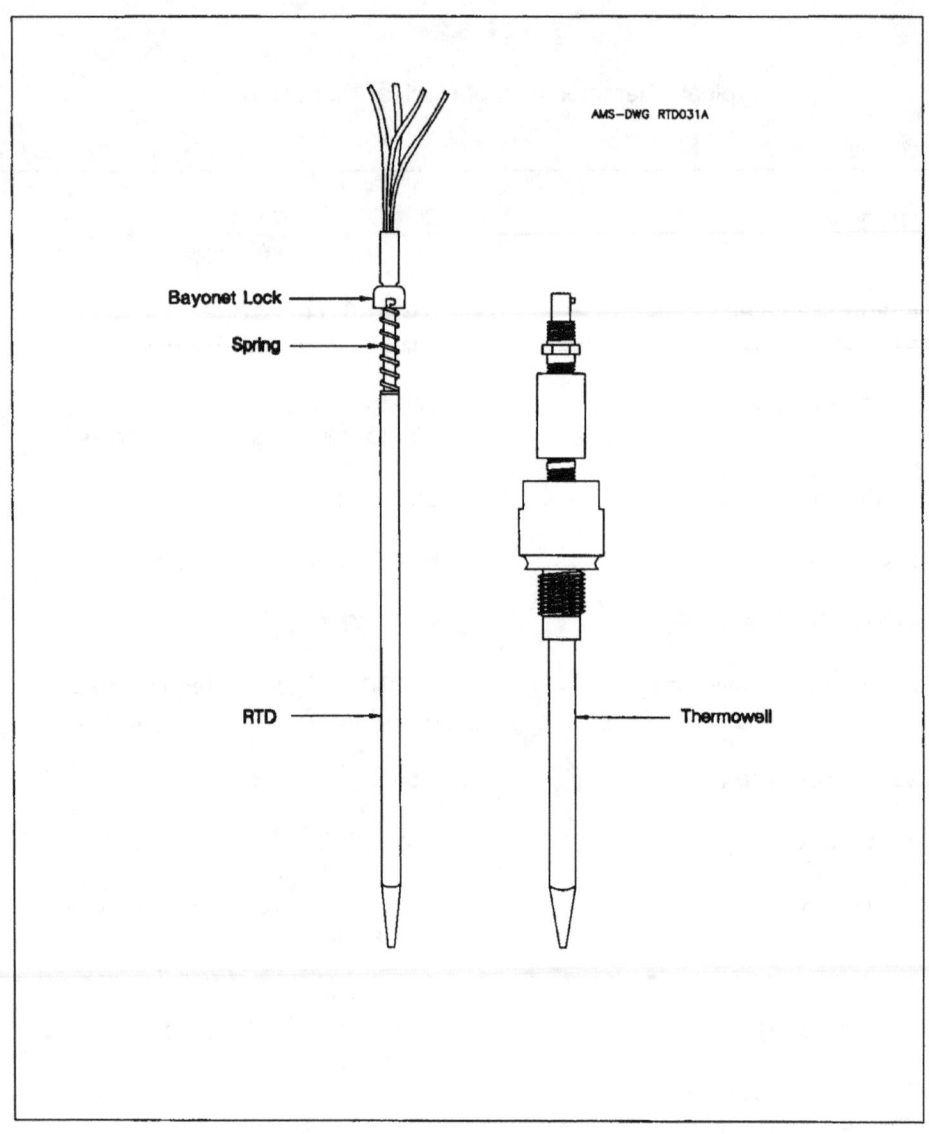

AMS—DWG RTD031A

Bayonet Lock

Spring

RTD

Thermowell

Figure 5-5. Illustration of a Spring-Loaded Tapered-Tip RTD.

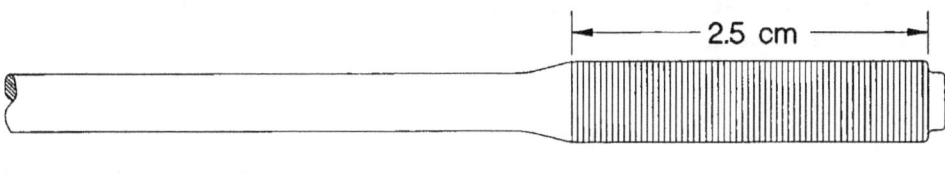

Silver Brazed

Silver Plugged

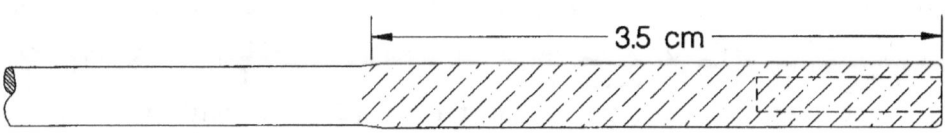

Silver Plated

Figure 5-6. Various Designs of Flat-Tip Thermowell-Mounted RTDs.

6. DEFINING RTD PERFORMANCE

The performance of an RTD is characterized by its accuracy and response time. Accuracy is a measure of how well the RTD may indicate a static temperature and response time defines how quickly the RTD may detect a temperature change. The accuracy and response time of an RTD are generally independent.

The deterioration of accuracy is called calibration drift or calibration shift. The deterioration of response time is called response time degradation. Accuracy can be restored by recalibration if the RTD is stable, but response time is an intrinsic characteristic that cannot be altered once the RTD is manufactured. In the case of thermowell-mounted RTDs, however, response time degradation due to movements of the RTD in its thermowell can sometimes be reversed.

Accuracy, uncertainty, and error are used interchangeably to describe the difference that may exist between the actual temperature of the process and the temperature indicated by the RTD. Since the actual temperature is not known, uncertainty is the most appropriate term. However, accuracy and error are more commonly used.

Accuracy is sometimes expressed in terms of a percentage of reading or percentage of span, but a more appropriate and unambiguous expression is to specify the accuracy in terms of a value at a given temperature or over a specific temperature range. For example, it is best to specify that the accuracy of an RTD is 0.2°C at 200°C. Another appropriate expression would be to specify, for example, that the accuracy of the RTD is 0.3°C for the temperature range of 0 to 300°C. The latter means that the maximum steady state difference that may exist between the temperature indicated by the RTD and the actual temperature being measured is within ± 0.3°C anywhere in the 0 to 300°C range.

The response time of an RTD is characterized by its time constant. This is defined as the time required for the RTD output to reach 63.2 percent of its final value following a step change in process temperature. Although this definition is meaningful only for first order dynamic systems, it is conventionally used to describe the response time of RTDs even though RTDs are not necessarily first order. This does not mean that a first order time constant can be reported as the overall time constant of an RTD. The overall time constant of an RTD must include the effect of all modal time constants. The number of modal time constants corresponds to the dynamic order of the RTD system.

RTD manufacturers usually specify the generic accuracy, stability, and response time of RTDs at a reference condition. While useful for comparative evaluation and selection of RTDs, this information may have little bearing on the actual performance achieved in an operating plant. The in-service performance depends not only on the as-built characteristics of the RTD, but also on the installation details, aging characteristics, and the process and environmental conditions.

7. MEASUREMENT OF RTD PERFORMANCE

This section presents a discussion on how an RTD is calibrated or response time tested in a laboratory or in a plant at operating conditions. The latter is referred to as in-situ testing, remote testing, or on-line testing.

7.1 Laboratory Calibration

Laboratory calibration of an RTD involves measuring its resistance at several known temperatures which are measured with an SPRT. Each pair of resistance versus temperature data is referred to as a calibration point. Since the resistance versus temperature curve of an RTD slightly departs from a straight line and resembles a parabola, at least three calibration points covering the range over which the RTD is used are necessary. The measured resistance versus temperature data are fit to a polynomial to provide calibration data for the temperatures in between the calibration points. A commonly used polynomial for temperatures above 0°C is the Callendar equation:

$$R(T) = R_0 \left[1 + \alpha \left(T - \delta \left(\frac{T}{100°C} \right) \left(\frac{T}{100°C} - 1 \right) \right) \right] \qquad (7.1)$$

Where:

T	:	Temperature in °C
R_0	:	Resistance at ice point (0°C)
α	:	Constant ($\Omega/\Omega/°C$)
δ	:	Constant (°C)
$R(T)$	:	Resistance at any temperature T

Alpha (α) is the average temperature coefficient of resistance over the 0 to 100°C interval. This coefficient is an indicator of purity of the platinum wire or the amount of strain present in the resistance element. Typical values of α are 0.003850 to 0.003925 $\Omega/\Omega/°C$.

Delta (δ) is the index of departure of the resistance versus temperature curve (R vs. T curve) from a straight line. The nominal value of δ for a platinum RTD is 1.5. A departure of more than 10 percent from the value of 1.5 for δ may be an indication of improper RTD calibration or a large calibration shift.

The fit of calibration data to the Callendar equation provides the calibration coefficients R_0, α, and δ. These coefficients define the resistance versus temperature relationship of the RTD for the temperature range in which the RTD is calibrated. If three points are used in the fitting, the resistance versus temperature curve will necessarily pass through the calibration points. If more than three points are used, then the fit is a least-squares solution of an over-determined algebraic system and the solution is the best least-squares fit to the data. In this case, the fit does not necessarily pass through the calibration points. Rather, the fit minimizes the sum of the squares of the differences between the measured points and the fit.

The Callendar equation is one of the many forms of polynomials used as an interpolation equation for RTDs. A general form is:

$$R(T) = R_0 (1 + AT + BT^2) \qquad (7.2)$$

Where:

$$A = \alpha \left(1 + \frac{\delta}{100} \right) \qquad (7.3)$$

$$B = - \, \alpha\delta / 10^4 \qquad (7.4)$$

In general, a polynomial of the following form may be used:

$$R(T) = R_0 (1 + AT + BT^2 + CT^3 + \dots) \qquad (7.5)$$

Although higher order polynomials provide for more accurate interpolation, the benefits in calibration of industrial RTDs may not be adequate for the effort involved in obtaining the additional calibration points needed for higher order polynomials.

The results of a laboratory calibration of an RTD are usually reported in a calibration chart that shows the resistance of the RTD at a number of temperatures. Table 7.1 shows a typical calibration chart. (Note that an actual calibration chart would be printed in 1°C increments to facilitate interpolation.) A problem with some calibration charts is that they are extrapolated far

TABLE 7.1

A Typical RTD Calibration Table
(Abbreviated)

RTD Model #   : 16A
RTD Serial #   : 16A
Calibration Date: 10/10/89

MEASURED DATA

Temperature (°C)	Resistance (Ω)
0.0093	200.1808
100.2310	277.4178
200.4108	352.2053
300.4243	424.5236

CALCULATED CALIBRATION CONSTANTS

α (Alpha) = 0.0038494 $\Omega/\Omega/°C$
δ (Delta) = 1.5413
R(0)   = 200.1768 Ω

CALIBRATION CHART

T (°C)	R(Ω)	T (°C)	R(Ω)	T (°C)	R(Ω)	T (°C)	R(Ω)
0.0	200.177	80.0	262.012	160.0	322.326	240.0	381.120
5.0	204.086	85.0	265.826	165.0	326.045	245.0	384.745
10.0	207.989	90.0	269.634	170.0	329.759	250.0	388.363
15.0	211.887	95.0	273.436	175.0	333.466	255.0	391.975
20.0	215.778	100.0	277.233	180.0	337.167	260.0	395.581
25.0	219.663	105.0	281.023	185.0	340.863	265.0	399.182
30.0	223.543	110.0	284.808	190.0	344.552	270.0	402.776
35.0	227.417	115.0	288.586	195.0	348.236	275.0	406.365
40.0	231.284	120.0	292.359	200.0	351.913	280.0	409.948
45.0	235.146	125.0	296.126	205.0	355.585	285.0	413.524
50.0	239.002	130.0	299.886	210.0	359.251	290.0	417.095
55.0	242.851	135.0	303.641	215.0	362.910	295.0	420.660
60.0	246.695	140.0	307.390	220.0	366.564	300.0	424.219
65.0	250.533	145.0	311.133	225.0	370.212		
70.0	254.365	150.0	314.870	230.0	373.854		
75.0	258.191	155.0	318.601	235.0	377.490		

beyond the highest temperature at which the RTD is calibrated. Any extrapolation is subject to much larger uncertainties than the interpolation uncertainty. This is discussed in Section 17.

7.2   Laboratory Response Time Testing

The laboratory response time of an RTD is measured in a rotating tank of water at 1 meter per second (m/s) flow rate. The test can be performed by warming or cooling the RTD in air above the water and plunging it into room temperature water or plunging from room temperature air into warm water. The former procedure is better because it eliminates problems with maintaining temperature stability in the rotating tank.

The RTD output is recorded when the RTD is plunged into the water. Two data channels are used, one for timing to identify when the RTD touches the water and one for the RTD output transient. The RTD output transient is recorded until it reaches steady state. The time constant of the RTD is directly measured from this transient by measuring the time for the RTD output to reach 63.2 percent of its final steady state value.

Since an RTD's response time depends on fluid velocity and temperature, laboratory response time results are useful only for comparative evaluation of RTDs and have very little bearing on response time results at process operating conditions. In thermowell-mounted RTDs, in addition to fluid velocity and temperature, the response time is strongly dependent on the fit between the RTD and thermowell. Therefore, if an RTD is tested in a laboratory inside a thermowell and subsequently installed into a plant and tested in another thermowell, there may be a large difference between the results of the two tests.

7.3 In-Situ Calibration

In-situ calibration of RTDs is possible by cross calibration tests which include one or more newly calibrated RTDs to be used as reference. If new RTDs are not included, the cross calibration is a test of consistency rather than accuracy. Cross calibration can be performed at a temperature plateau to provide a one-point check or at several temperature plateaus to provide a narrow range calibration. Cross calibration can also be done when the plant is undergoing a temperature ramp as long as the ramp rate is constant and slow enough compared to the response time of the RTDs. A detailed discussion of the cross calibration method is presented in Section 23.

The Johnson noise method that has been under development for in-situ calibration of RTDs cannot be used remotely because it cannot resolve differences of less than a few degrees celsius at the end of a few hundred feet of wire.[3] Johnson noise is the electronic noise that occurs in any conductor as a result of thermal excitation of the conduction electrons. This creates a noise voltage (in the nanovolt range at moderate temperatures) which is directly related to temperature.

Properties of the conductor do not affect the noise voltage, so the measurement is independent of the sensor material composition.

7.4 In-Situ Response Time Testing

In-situ response time testing of RTDs is readily done using the Loop Current Step Response (LCSR) method. This method has been approved by the NRC for use in nuclear power plants. The test involves applying a small electric current (40 to 80 milliamps) to the RTD leads to cause internal heating in the platinum sensing element. The current can be applied at the end of extension wires several hundred feet away from the RTD. The extension wires do not heat up appreciably because of their low resistance, but the RTD sensing element heats up and causes a temperature transient in the RTD that settles at about 5 to 10°C above the fluid temperature around the RTD. This transient is recorded and analyzed to give the RTD time constant. The time constant obtained from the LCSR test is within less than ten percent of the time constant which would be obtained if the process temperature experienced a step change. The LCSR test is performed remotely at normal operating conditions. Therefore, it provides the actual in-service response time of the RTDs and accounts for the effects of process conditions, installation, and aging. The details of the LCSR technology are available from a number of reports published by the Electric Power Research Institute.

8. DEFINITION OF AGING

The term aging as used in this report refers to decalibration or response time degradation of RTDs with time in normal environments and under normal operating conditions in Pressurized Water Reactors (PWRs). This definition is based on NPAR's definition of aging which is, "the cumulative degradation that occurs with the passage of time in a component, system, or structure which can, if unchecked, lead to loss of function and impairment of safety". Since the performance of RTDs is tested periodically, the degradation does not accumulate. Therefore, the word cumulative was deleted in our definition. Furthermore, we concentrated on the aging that occurs in an 18 month period, the length of a typical PWR fuel cycle and the period of time between periodic response time and cross calibration tests currently performed in nuclear power plants.

The normal environments for nuclear plant RTDs are heat, humidity, vibration, temperature cycling, and mechanical shock. These are referred to as normal stressors. One or more of these stressors is always acting on the RTD whether it is in storage or installed in a plant at operating or shutdown conditions. Although these stressors result in aging, RTDs are designed to survive them during their quantified life except for unanticipated failures.

Nuclear plant RTDs usually experience additional stress resulting from handling, installation, maintenance, and design or manufacturing flaws. These are referred to as abnormal stressors. Generally, aging that results from abnormal stressors is difficult to quantify because the magnitude of abnormal stressors themselves can not be quantified. In lieu of testing to study the effects of abnormal stressors, a series of tests referred to as "destructive tests" were performed as reported in Appendix A. In these tests, we simulated mishandling, exposed representative RTDs to high temperature and high humidity, and performed abnormal temperature cycling. These tests were intended to provide an understanding of the tolerance level of the RTDs to abnormal conditions.

9. AGING OF RTDs

Normal aging of RTDs occurs from long-term exposure to any combination of heat, humidity, vibration, temperature cycling, and mechanical shock. Nuclear radiation can also affect RTD performance, but this was not studied here as it was beyond the scope of the project. Since primary coolant RTDs are remote from the reactor core, they are normally unaffected by nuclear radiation except for gamma which may cause degradation in the insulation and other RTD materials.

The normal conditions to which primary coolant RTDs are typically exposed in a PWR are summarized in Table 9.1.

Figure 9.1 illustrates the components of a typical RTD. The sensing element and extension wires are sealed in a metallic sheath and held in place and insulated from the sheath by insulation material in the form of a powder or cement. All components of the RTD are subject to aging. Aging of the sensing element affects calibration and aging of the insulation material affects both the calibration and response time. Aging of extension wires and the seal are important only when they progress to the point of affecting the sensing element or the insulation material. For example, seal aging is of no concern until it has progressed to the point where the seal can not keep moisture from entering the RTD. Normal aging of the sheath does not have a significant effect on RTD performance and is therefore not of concern.

The potential effects of normal stressors are summarized in Table 9.2 and discussed below.

9.1 Aging Effects on Calibration

A significant calibration shift should not occur in an RTD as long as the sensing element is not stressed or contaminated after calibration and the insulation material is kept in place and dry. Any new stress, contamination, or metallurgical changes in the sensing element or moisture in the insulation material can cause a calibration shift.

Stress results from any combination of heat, vibration, temperature cycling, and mechanical shock. The effect of temperature is the most important. This is because the RTD materials have different thermal expansion coefficients causing the element to experience stress whenever the temperature changes. The resistance of the sensing element increases with tension stresses and decreases with compression stresses. For small temperature variations, the stress reverses itself but for large ones, the effect is not reversible except by annealing. Chemical contamination and oxidation of the sensing element results from long-term exposure to high temperatures. To avoid oxidation, RTDs may be built with reducing atmosphere in the sheath. However, this leads to

TABLE 9.1

Normal Aging Conditions for Primary
Coolant RTDs in PWRs

Temperature Range	300 to 320°C
Temperature Cycling Conditions	Shutdowns, Start-ups, Plant Trips
Temperature Fluctuations	± 0.5°C
Containment Temperature Range	50 to 60°C
Storage Temperature	Ambient Temperature (approx. 20°C)
Containment Humidity Range	10 to 90%
Vibration Sources	Flow Induced Vibration Vibration of Nearby Machinery
Sources of Mechanical Shock	Shock in Shipping, Handling, Installation, and Plant Trips

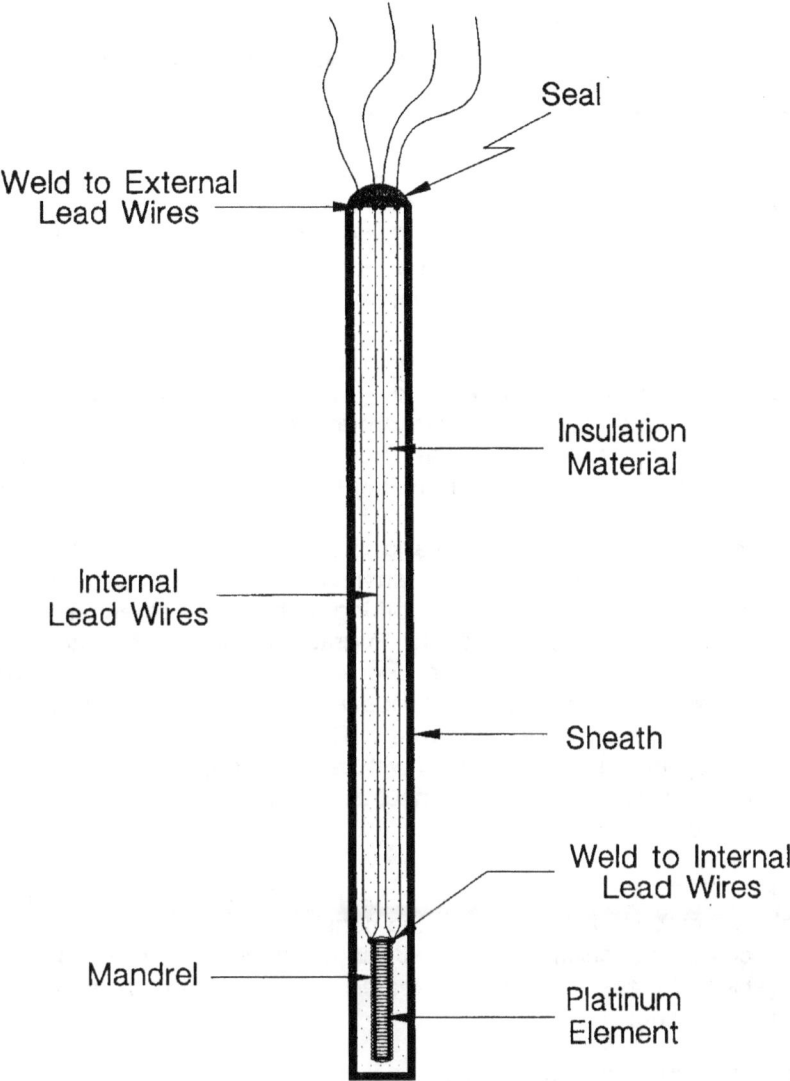

Seal

Weld to External
Lead Wires

Insulation
Material

Internal
Lead Wires

Sheath

Weld to Internal
Lead Wires

Mandrel

Platinum
Element

Figure 9-1. Components of a Typical RTD.

TABLE 9.2

Potential Effects of Normal Stressors on RTD Performance

HEAT. Long-term exposure to high temperature affects material properties. Chemical contamination or metallurgical changes occur in the platinum element and cause a calibration shift. Gaps and cracks develop in the insulation material and result in response time degradation. The RTD seal may dry out, shrink, or crack and allow moisture into the sheath.

HUMIDITY. The humidity levels inside reactor containment are in the range of 10 to 90 percent. Some moisture can leak into the RTD with long-term exposure to the temperatures that exist around the head of the RTD. Moisture in the RTD reduces the insulation resistance and causes calibration error. In addition to causing calibration error, moisture results in a noisy RTD output.

VIBRATION. Flow induced vibration and vibration generated by nearby machinery during plant operation are transmitted to RTDs through the reactor piping system. This may cause cold working of the RTD element and result in a calibration shift or can cause the RTD to gradually move out of the thermowell and result in response time increases. If the RTD is spring loaded in the thermowell, vibration can cause loosening of the spring and allow the RTD to change position in the thermowell. The consequence will normally be changes in response time. Combined with temperature, vibration can cause displacement or redistribution of the insulation materials and result in response time degradation.

TEMPERATURE CYCLING. This causes expansion and contraction of sensor materials and may result in stress on the sensing element. Any stress on the element can cause a calibration shift. Changes in response time can result due to gaps and cracks that can be created in the insulation materials from temperature cycling.

MECHANICAL SHOCK. Any shock to the RTD from sudden changes in plant operating conditions can cause degradation in the same manner as vibration.

contamination due to migration of metal ions from the sheath to the sensing element at temperatures above 500°C.

Metallurgical changes such as grain growth occur at temperatures above 420°C. Cold working (or work hardening) results from vibration and mechanical shock and can be eliminated by annealing, which requires heating the RTD above 400°C.

The insulation resistance of an RTD decreases as moisture enters the sheath. The electrical resistance of an RTD is a parallel combination of two resistances: the sensing element and the insulation resistance (Figure 9.2). The insulation resistance is normally high compared to that of the sensing element and has a negligible effect on resistance measurement. However, with moisture, the insulation resistance decreases and causes the RTD to indicate a lower temperature than normal. The insulation resistance is measured with an instrument (called a megohm meter) which applies 50 to 100 volts DC across the insulation between any RTD wire and the sheath. It is important to point out that insulation resistance measurements are often difficult to make if there is much moisture in the RTD. More specifically, the megohm meter will not remain stable enough to make a reliable measurement. To overcome this problem, most procedures give a specific duration for the measurement. Minor insulation resistance problems can be corrected by heating to drive the moisture out. However, if the moisture has entered the RTD due to seal degradation, the correction will not hold for a long time.

In-situ measurement of insulation resistance of installed RTDs may provide information about the integrity of the seal and the insulation material. However, the results of such measurements must be used with caution as the insulation resistance values may be dominated by the insulation properties of extension cables or connectors.

At high temperatures, moisture in the RTD is not normally a major concern because water vapor is likely to diffuse out of the RTD. However, since at high temperatures the insulation resistance significantly decreases, any remaining moisture in the RTD may have a significant impact on the insulation resistance value. Figure 9.3 shows the insulation resistance of dry magnesium oxide (MgO) as a function of temperature. Note that the insulation resistance decreases by an order of magnitude for every 100°C increase in temperature. Magnesium oxide is used for support and insulation of sensing elements in industrial or commercial grade RTDs. In nuclear grade RTDs, the insulation is usually made of special material to provide good and stable insulation resistance without resulting in a large response time.

9.2 Aging Effects on Response Time

Response time degradation results from changes in the heat transfer properties of the insulation material. Gaps and cracks which may develop in the insulation materials from

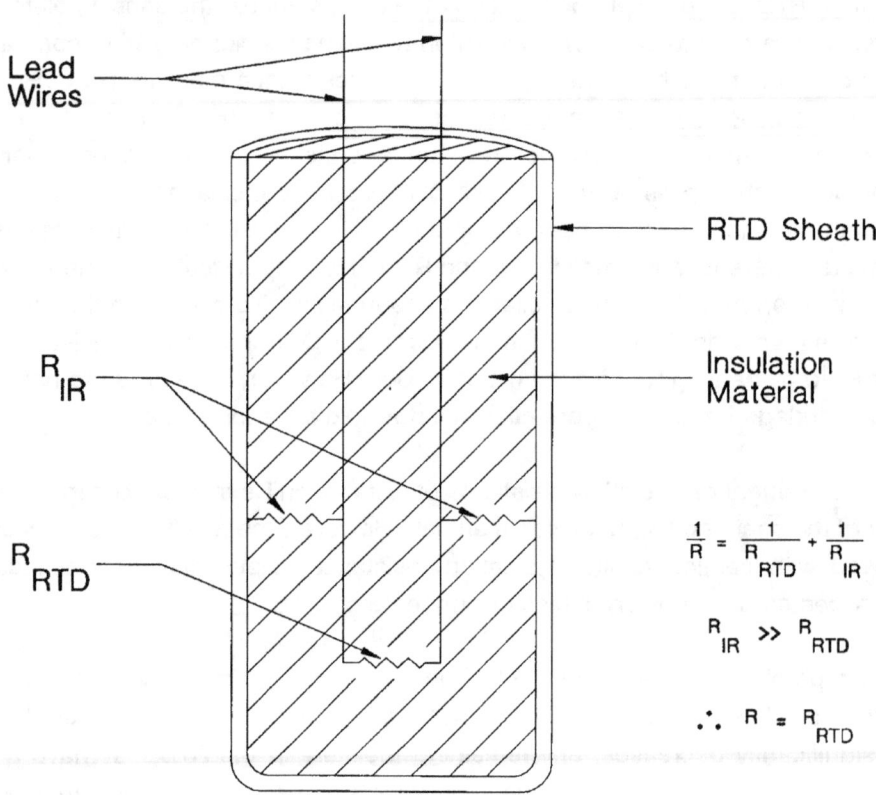

Figure 9-2. Electrical Resistances of an RTD.

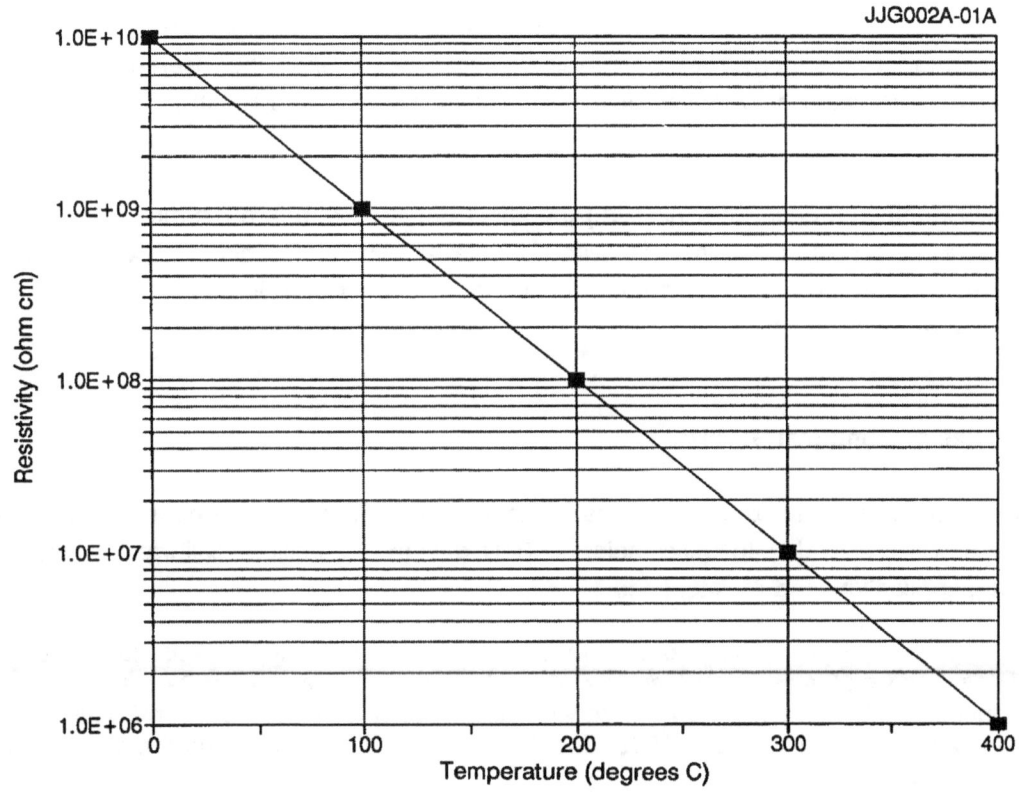

Figure 9-3. Insulation Resistance of MgO as a Function of Temperature.

long-term exposure to high temperature, vibration, and temperature cycling are detrimental to dynamic performance. If moisture enters the RTD, the response time may decrease at the cost of a calibration shift. Although improvement in response time with age is possible, an RTD whose response time continues to decrease with age could be suffering from degradation of insulation resistance.

A major cause of response time degradation in nuclear plant RTDs is the change that occurs in the RTD/thermowell interface in well-mounted RTDs. Experience has shown that air gaps (Figure 9.4) in the RTD/thermowell interface play a major role in controlling the overall response time of the RTD. Changes of as little as a few hundredths of a millimeter in the size of the air gap caused by vibration, shock, and other mechanical effects during plant operation, installation, handling, or dimensional tolerances will change the response time significantly. If the RTD is spring loaded into the thermowell, mechanical effects may change the insertion length or the contact pressure, increase the size of the air gap in the well, and result in a response time increase.

A reasonable approach for minimizing the effect of the air gap in a thermowell has been to electroplate the sensing tip of the RTD with a thin layer of silver or gold (Figure 9.5). Another approach involving the use of a thermal compound in the thermowell has not been successful as a long-term solution.

9.3 Effects of Abnormal Stressors

This section provides examples of performance problems that have been encountered in nuclear power plants due to deficiencies in RTD design, fabrication, calibration, installation, handling, application, and maintenance.

- Design. Design problems in both direct immersion and well-type RTDs have caused performance problems in nuclear power plants. Direct immersion RTDs that are short cannot be calibrated accurately due to stem loss problems. Also, there have been many failures of direct immersion RTDs due to failure of the seal around the lead-wire penetration.

- Fabrication. In one nuclear plant, about fifty percent of newly installed RTDs failed at start-up, presumably due to fabrication problems. The problem did not repeat itself when the failed RTDs were replaced with another batch of RTDs.

  Tapered-tip RTDs have been found to have problems with removeability from the thermowell. Some RTDs have been destroyed in the process of removing them from a thermowell. The problem is sometimes in the weld where the tapered section is attached to the stem.

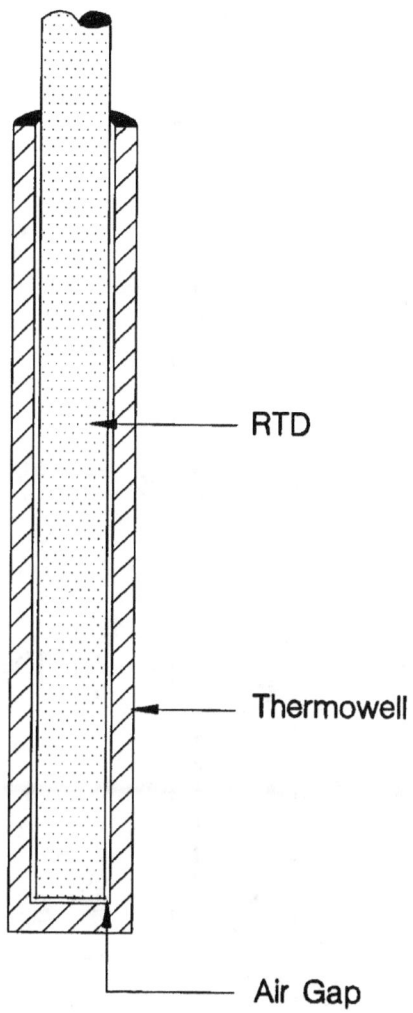

RTD

Thermowell

Air Gap

Figure 9-4. Illustration of Air Gap in RTD/Thermowell Interface.

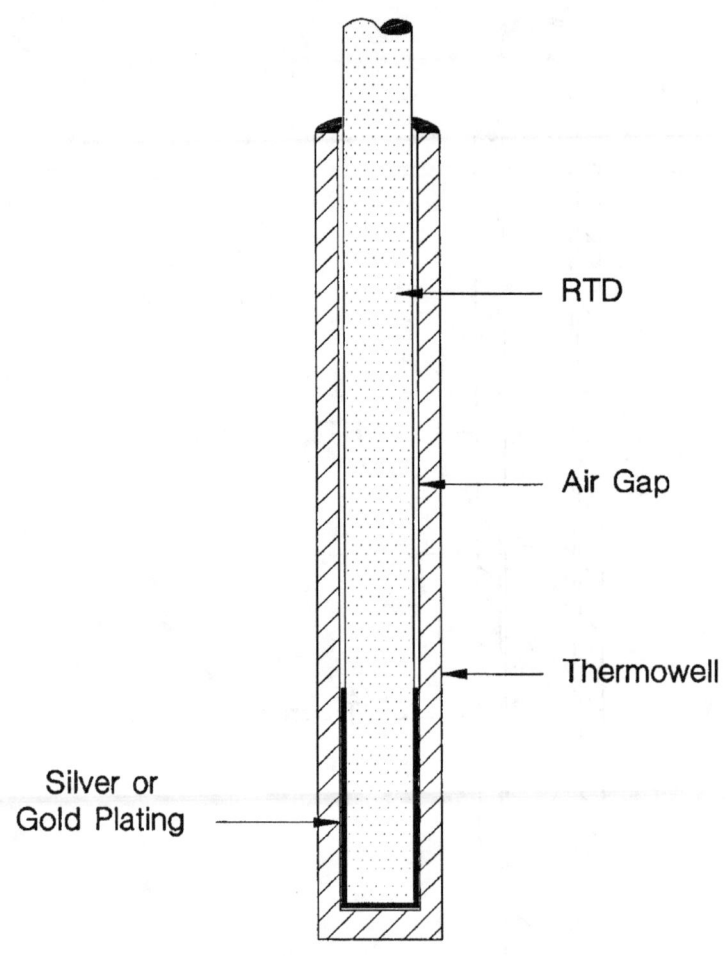

Figure 9-5. Illustration of a Silver or Gold Plated RTD in Thermowell.

Thermowells have been found to have been drilled deeper or larger in diameter than intended. This problem causes a large response time and cannot be resolved except by using a thermal coupling compound in the thermowell or by replacing the thermowell.

- Improper Calibration Tables. Nuclear plant RTDs are supplied with calibration tables generated by the manufacturer. A few cases of significant temperature errors have been traced to improper factory calibration. In one case, a computer-aided calibration of several RTDs had not allowed the RTDs to come to equilibrium with the bath before measurements were made. As a result, the RTDs indicated lower than the actual temperatures. In plants where the accuracy of RTDs is tested by cross calibration, a common mode error such as this is detrimental to safety because it can go undetected for a long time.

- Installation. Proper installation of a well-type RTD into the thermowell is crucial in achieving optimum response time. Numerous cases of response time failures have occurred due to inadequate insertion of the RTD into the thermowell. These problems have occurred because of personnel errors or problems in the dimensional tolerances of RTDs and thermowells or in the way that the RTD is secured in the thermowell. In one plant, 30 percent of reactor coolant RTDs were found with unacceptable response times. One of these had a response time of 37 seconds instead of the 6 seconds that was required. This resulted from improper seating of the RTDs into their thermowells. The plant had to be shut down to restore the RTD response times.

- Storage and Handling. RTDs are delicate instruments which must be handled properly. If an RTD is dropped or hit against a hard object, the calibration may be affected. If an RTD is stored in a humid environment, it may suffer a calibration shift. Mishandling in shipment has been the cause of several failures of nuclear plant RTDs.

- Maintenance. RTDs are sometimes removed and recalibrated. Every time an RTD is removed, there is potential for damage in handling and installation. Removal for recalibration should be done only after comparing the benefit with the potential for damage to the RTD. It is not advisable to adopt periodic removal and recalibration of RTDs as a means of maintaining accuracy requirements. It is better to perform a screening test such as cross calibration to identify and remove only those RTDs that do not meet the requirement. This approach is consistent with the signal validation concept and "calibration reduction" philosophy whereby a simple test is conducted to segregate the sensors that require maintenance.

A major consequence of removing an RTD is changes in response time that may occur every time an RTD is removed from its thermowell. This would require performing a response time test after the RTD is returned to its thermowell to verify that the RTD is properly reseated in the thermowell and have the desired response time.

10. DESCRIPTION OF WORK

The work reported herein involved testing 51 nuclear grade and 17 commercial grade RTDs. The nuclear group included 21 dual element RTDs for a total of 72 elements. A listing of these RTDs is given in Table 10.1 and a picture is shown in Figure 10.1. This is followed by the listing of the commercial grade RTDs in Table 10.2. The commercial grade RTDs were included in the project for comparison purposes and for the study of failure modes. The tag numbers shown in these tables were assigned in this project to help in presenting the results. Most of the RTDs used in the project were well-type, but thermowells were not included in the aging tests as they do not generally affect the drift characteristics of RTDs.

The nuclear grade RTDs listed in Table 10.1 include eight from the Phase I project, 12 provided by manufacturers, and 31 provided by interested utilities. These RTDs represent at least six from each of the four U.S. manufacturers of nuclear grade RTDs. The RTDs provided by manufacturers and the commercial grade RTDs which were purchased for the project are new. The rest included both new and old RTDs that have been used in other projects and in some nuclear power plants. Note that all nuclear grade RTDs studied here are of the type used for narrow range temperature measurements in the safety system of PWRs. Although similar RTDs are used in wide range temperature measurements, the accuracy requirements for wide range RTDs are not as stringent.

The project was conducted from October 1987 to March 1990. In the first four months, the tests were performed using the Phase I RTDs while waiting for new RTDs to arrive. During this period, an automatic RTD calibration and monitoring system was developed and a thermal aging station utilizing two tube furnaces was set up. The RTD monitoring system is illustrated in Figure 10.2. It consists of an electronic switching unit called a multiplexer or mux, a precision digital multimeter, and a data acquisition and control computer. The system was used to make the following measurements during the aging process.

- **Element Resistance**   Compensated measurement of the resistance of the sensing element.

- **Loop Resistance**   The resistance of the sensing element and the extension leads.

- **Insulation Resistance**   The resistance between any extension lead and the sheath.

- **Isolation Resistance**   The insulation resistance between the two elements of a dual RTD.

TABLE 10.1

Listing of Nuclear Grade RTDs
Tested in This Project

Item	Tag	Description

RTDs FROM PHASE I

Item	Tag	Description
1	03	Sngl 200Ω Well-type
2	04	Dual 200Ω Well-type
3	05	Dual 100Ω Well-type
4	07	Sngl 200Ω Well-type
5	08	Sngl 200Ω Well-type
6	09	Dual 200Ω Well-type
7	11	Dual 200Ω Well-type
8	12	Dual 200Ω Well-type

NEW RTDs PROVIDED BY MANUFACTURERS

Item	Tag	Description
9	13	Dual 200Ω Well-type
10	14	Sngl 200Ω Well-type
11	15	Dual 200Ω Well-type
12	16	Dual 200Ω Well-type
13	17	Dual 200Ω Well-type
14	18	Sngl 200Ω Well-type
15	19	Sngl 200Ω Well-type
16	20	Sngl 200Ω Well-type
17	21	Sngl 100Ω Well-type
18	22	Dual 200Ω Well-type
19	23	Sngl 200Ω Wet -type
20	24	Sngl 200Ω Wet -type

(continued on the next page)

TABLE 10.1
(continued)

Item	Tag	Description
		RTDs PROVIDED BY UTILITIES
21	56	Sngl 200Ω Well-type
22	57	Sngl 200Ω Well-type
23	58	Sngl 200Ω Well-type
24	59	Sngl 200Ω Well-type
25	60	Sngl 200Ω Well-type
26	61	Sngl 200Ω Well-type
27	62	Sngl 200Ω Well-type
28	63	Sngl 200Ω Well-type
29	64	Sngl 200Ω Well-type
30	65	Sngl 200Ω Well-type
31	66	Sngl 200Ω Well-type
32	67	Sngl 200Ω Well-type
33	68	Sngl 200Ω Well-type
34	69	Sngl 200Ω Well-type
35	70	Sngl 200Ω Well-type
36	97	Dual 200Ω Well-type
37	98	Dual 200Ω Well-type
38	99	Dual 200Ω Well-type
39	90	Sngl 200Ω Wet -type
40	91	Sngl 200Ω Wet -type
41	85	Sngl 200Ω Well-type
42	86	Sngl 200Ω Well-type
43	87	Sngl 200Ω Well-type
44	77	Dual 200Ω Well-type
45	78	Dual 200Ω Well-type
46	79	Dual 200Ω Well-type
47	80	Dual 200Ω Well-type
48	81	Dual 200Ω Well-type
49	82	Dual 200Ω Well-type
50	83	Dual 200Ω Well-type
51	84	Dual 200Ω Well-type

Sngl: Single element RTD
Dual: Dual element RTD

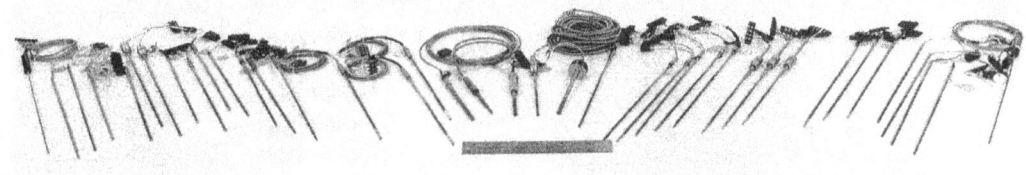

Figure 10-1.   Photograph of Representative Nuclear Grade RTDs
                Tested in This Project.

TABLE 10.2

Listing of Commercial Grade RTDs
Tested in This Project

Item	Tag	Description
1	25	Sngl 100Ω Well-type
2	26	Sngl 100Ω Well-type
3	27	Sngl 100Ω Well-type
4	28	Sngl 100Ω Well-type
5	35	Sngl 100Ω Well-type
6	36	Sngl 100Ω Well-type
7	37	Sngl 100Ω Well-type
8	38	Sngl 100Ω Well-type
9	42	Sngl 100Ω Well-type
10	43	Sngl 100Ω Well-type
11	44	Sngl 100Ω Well-type
12	51	Sngl 100Ω Well-type
13	52	Sngl 100Ω Well-type
14	53	Sngl 100Ω Well-type
15	54	Sngl 200Ω Well-type
16	40	Sngl 100Ω Well-type
17	95	Sngl 100Ω Wet -type

Sngl : Single element RTD
Dual : Dual element RTD

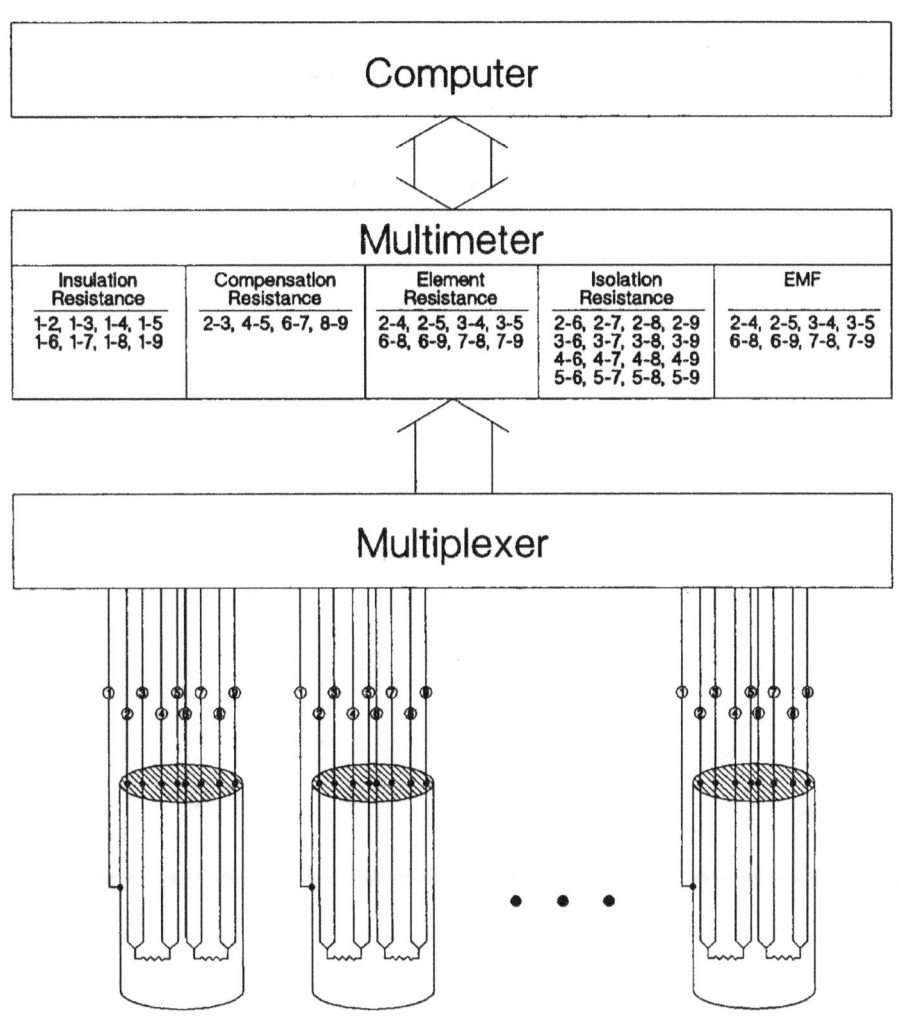

Figure 10-2. Illustration of RTD Monitoring System.

- **Open Circuit Voltage**   The EMF across the element.

- **Compensating Loop Resistance** Resistance of the two wires at one or both ends of the sensing element that are normally used for lead wire compensation.

These measurements were made once every hour on all RTDs being aged. The data were stored on computer disks and subsequently reviewed to identify any problems. Figure 10.3 shows the time histories of these measurements for a normal RTD. The same information is shown in Figure 10.4 for an unstable RTD. Note that there is not a good correlation between the time histories and the RTD instability. As such, the RTD monitoring results were used only for detecting major failures.

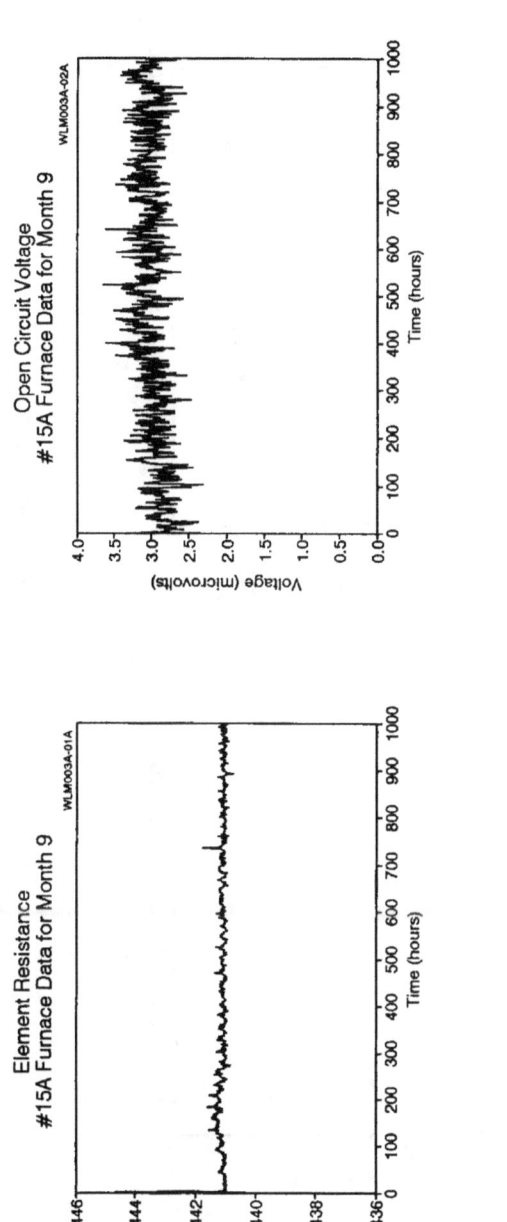

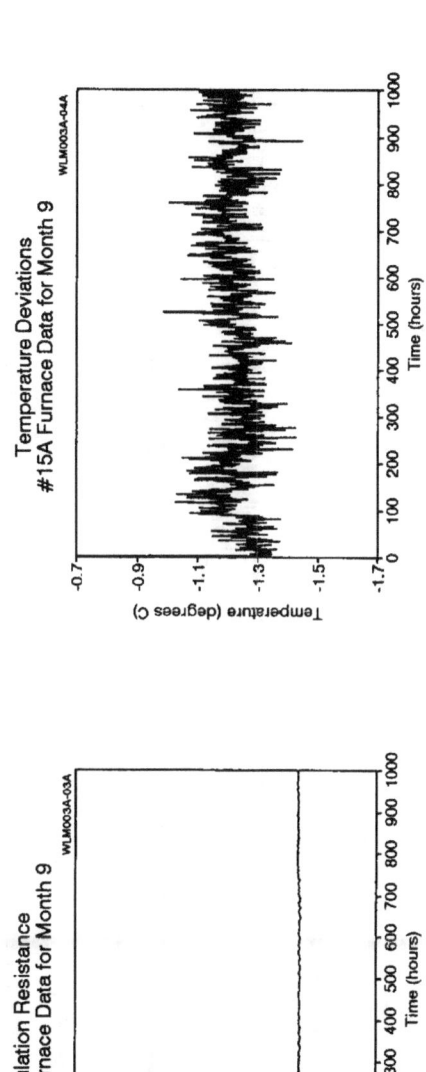

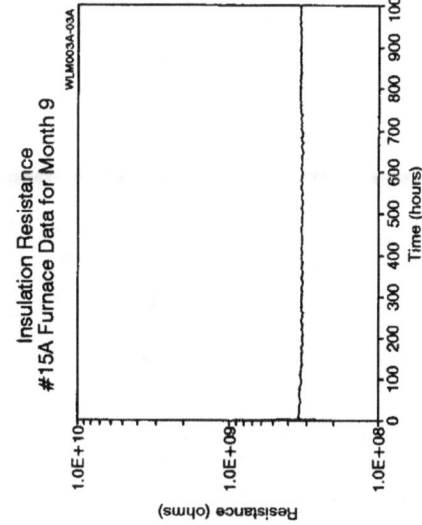

Figure 10-3. Time Histories for a Normal RTD.

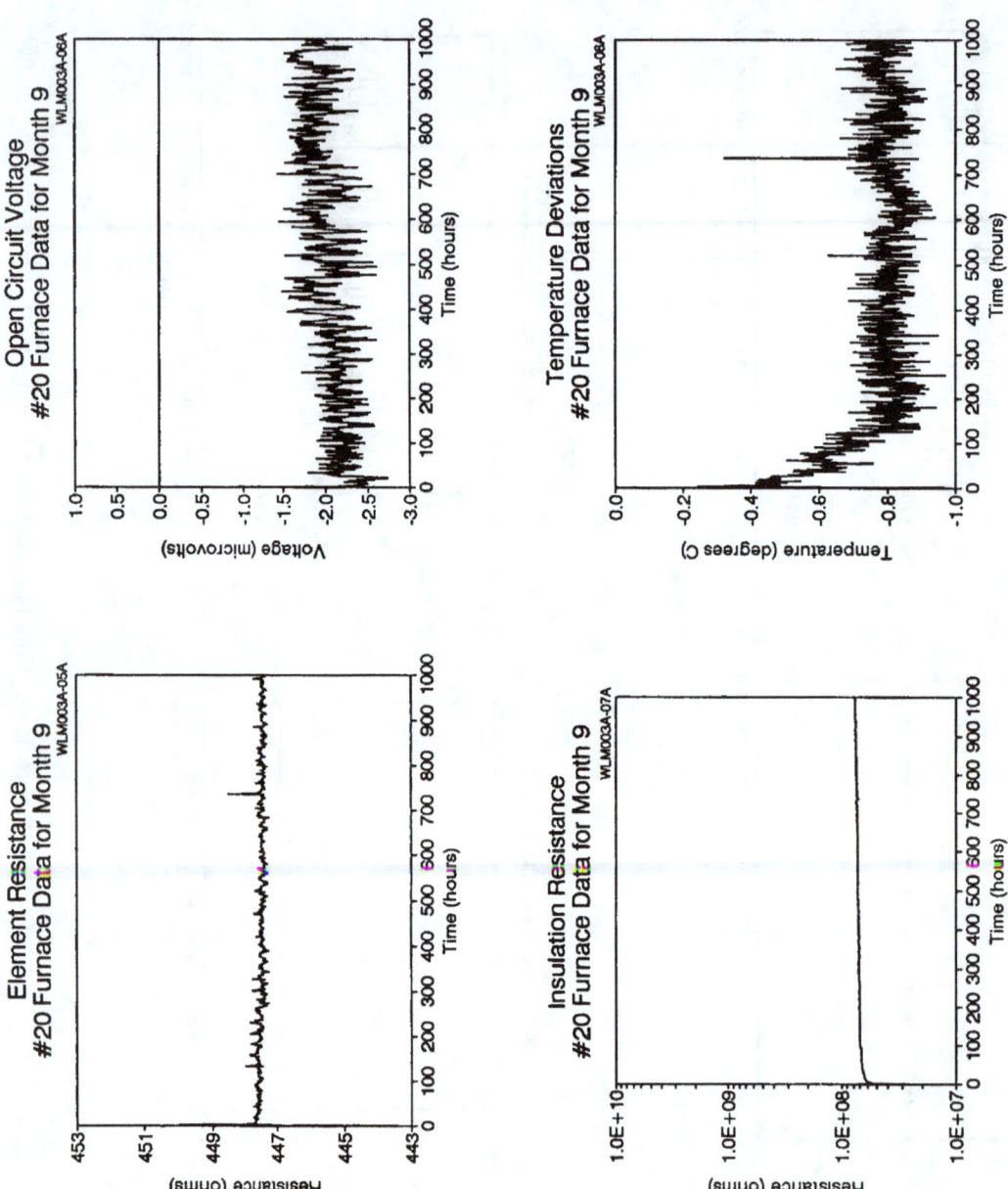

Figure 10-4. Time Histories for an Abnormal RTD.

- 44 -

11. CALIBRATION SYSTEM DEVELOPMENT

An important requirement for conducting this project was to develop a system for automatic, accurate, and repeatable calibration of RTDs. Most of the equipment and software to accomplish this was available at the beginning of the project from earlier developments including that of Phase I. These were integrated and improved to provide an accurate calibration system for the range of 0 to 300°C with minimum possibility for operator error.

The system consists of an ice bath, a well insulated stirred oil bath, an SPRT, an electronic switching system (multiplexer), a precision digital multimeter, and a computer for data acquisition and adjustment of oil bath temperature. The system is illustrated in Figure 11.1. A cylindrical copper block with seven holes, one at the center and six around the perimeter, is used in the bath for improved temperature stability and uniformity. Stability is a measure of peak to peak temperature fluctuations in the block and uniformity is a measure of spatial temperature distribution in the block. Figure 11.2 shows the bath stability at 300°C with and without the copper block. The stability was measured with an SPRT in the center of the bath. It is apparent that the copper block helps improve the stability by a factor of at least 3. An additional measure for minimizing the effect of bath instability is to make simultaneous measurements on the SPRT and the RTD being calibrated.

Temperature uniformity in the copper block was determined using two SPRTs, one in the center and one in the perimeter. The bath was allowed to stabilize for 30 minutes after which the outputs of the two SPRTs were monitored for one hour. A portion of the data for 300°C is shown in Figure 11.3. The difference between the average of the two traces is about 0.003°C. This experiment was performed in two identical oil baths with similar results.

For automated calibration, a computer program was written to adjust the oil bath temperature, determine when the bath was stable, and perform data acquisition. The computer routine for measuring bath stability uses continuous data from an SPRT and performs real-time standard deviation calculations until a predetermined stability criteria is met. At this point, 25 pairs of measurements are made on each RTD and the results averaged. Each pair includes the resistance of the SPRT and one of the RTDs being calibrated.

A picture of the calibration equipment developed is shown in Figure 11.4. Two oil baths and one ice bath are shown. Two oil baths were used in the project to accommodate testing of 14 RTDs at a time. For routine calibration of the RTDs during this research, four calibration points were used: 0, 100, 200, and 300°C. The data were fit to the Callendar equation (see Section 7) and the calibration constants ($\propto$, δ, and R_0) were identified. To compare two calibrations on the same RTD, the resistance versus temperature data for the two calibrations were subtracted and the results were plotted in terms of temperature differences over the range of 0 to 300°C. Typical difference plots are shown in Figure 11.5. Such plots were used to identify the maximum difference over the range of 0 to 300°C or the difference at a given temperature.

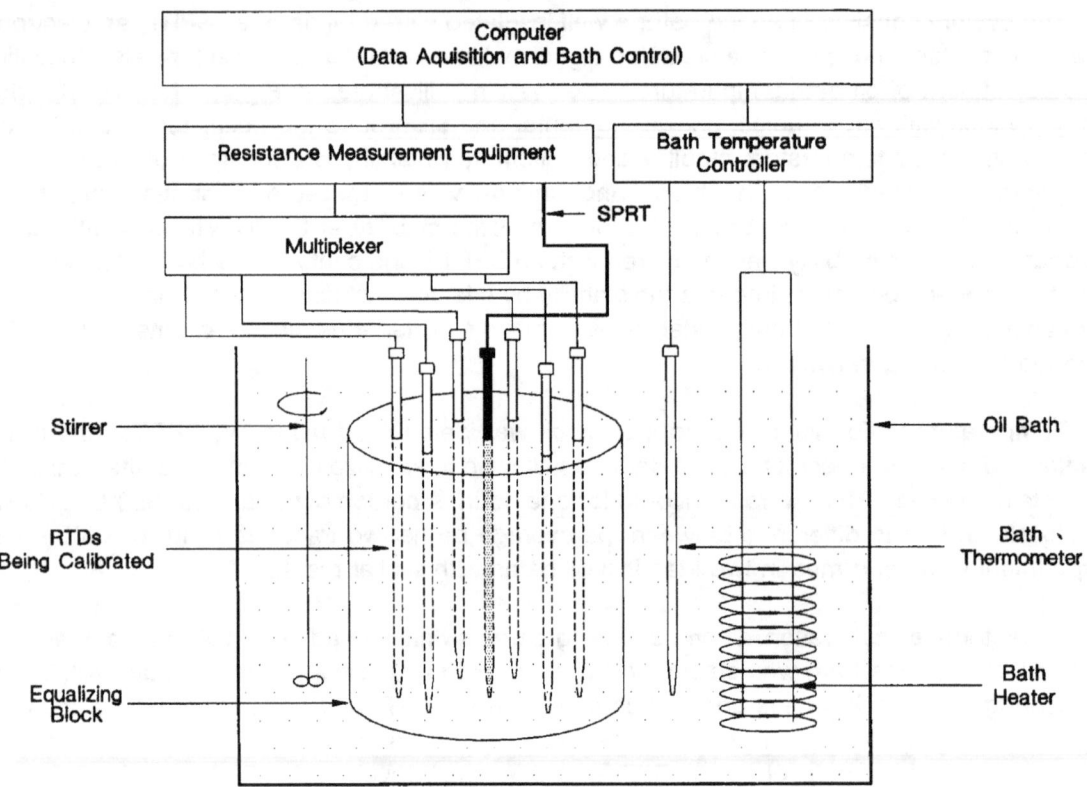

Figure 11-1. RTD Calibration System.

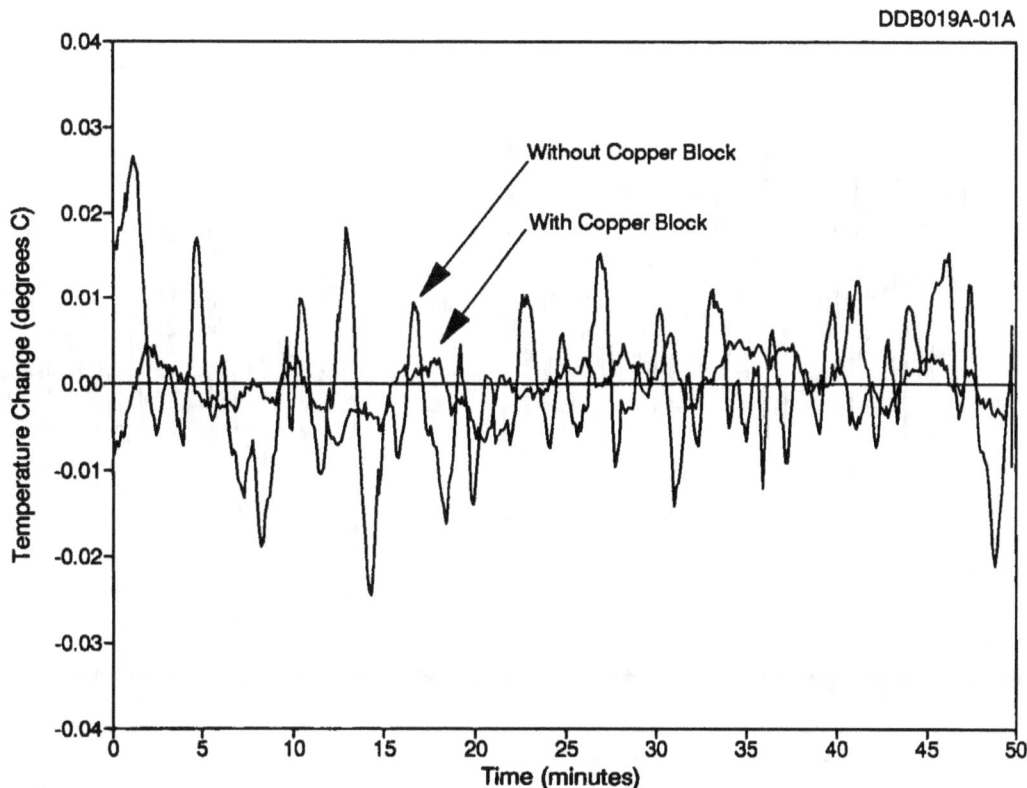

Figure 11-2.   Oil Bath Stability at 300°C With
and Without Copper Block.

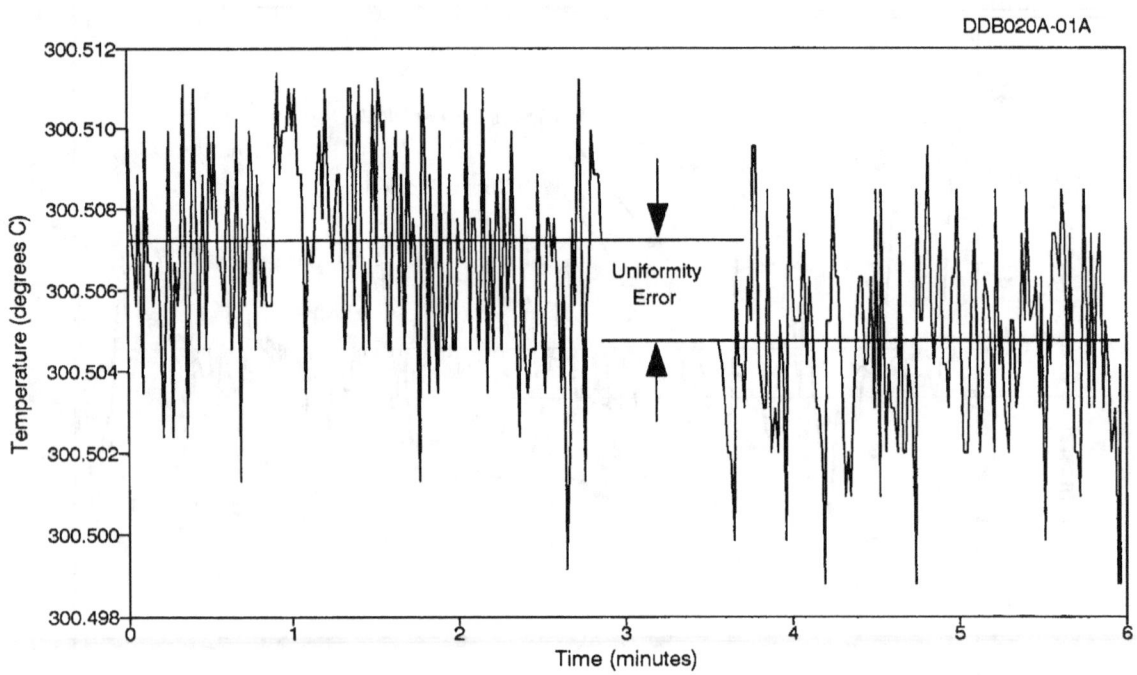

Figure 11-3.   Oil Bath Stability and Uniformity
                Data at 300°C.

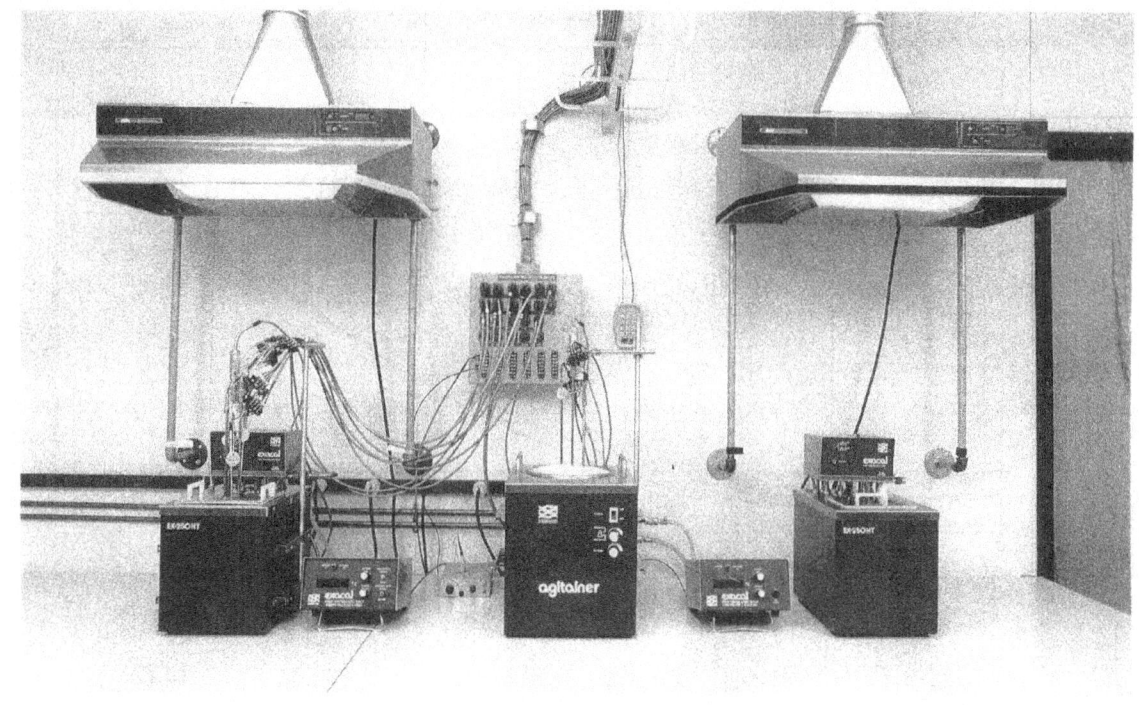

AMS—DWG CAL009A

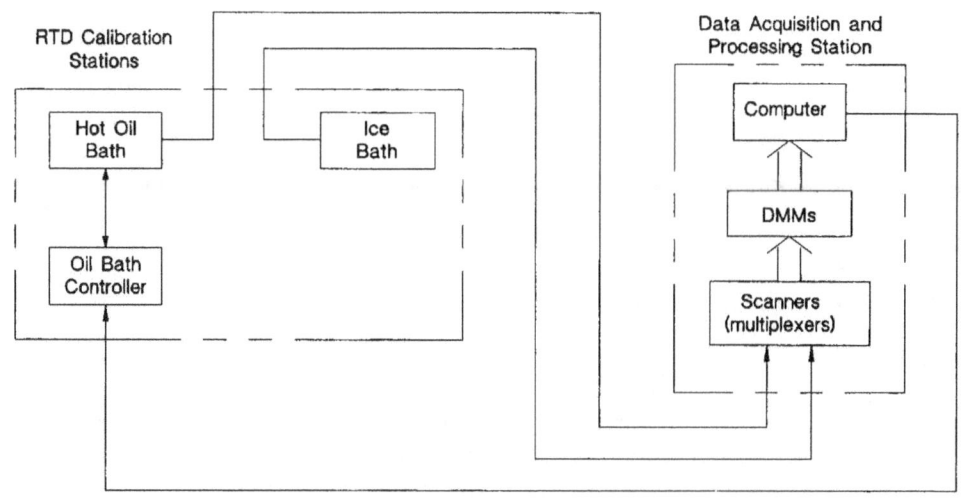

Figure 11-4. RTD Calibration Equipment.

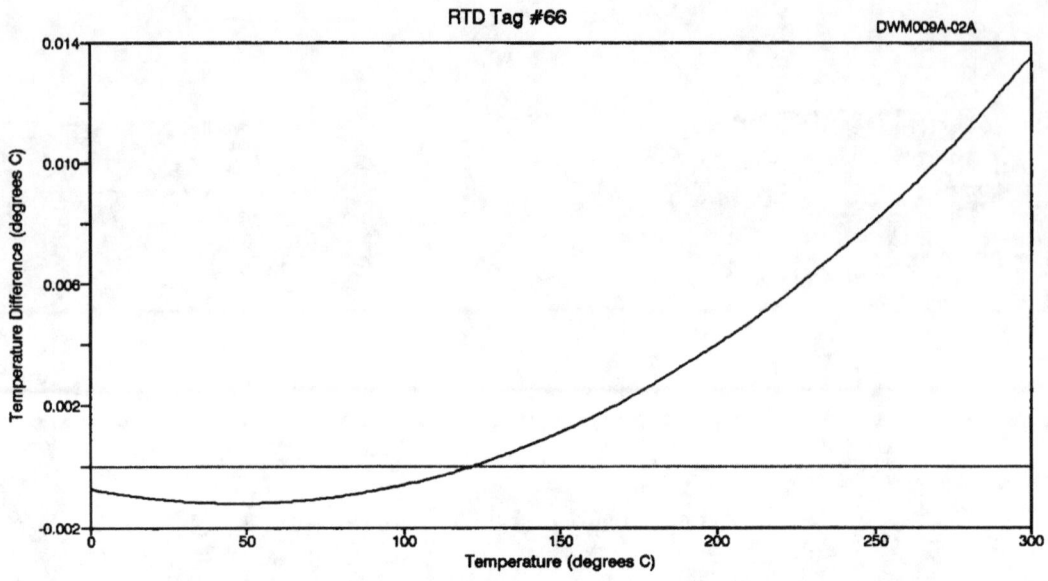

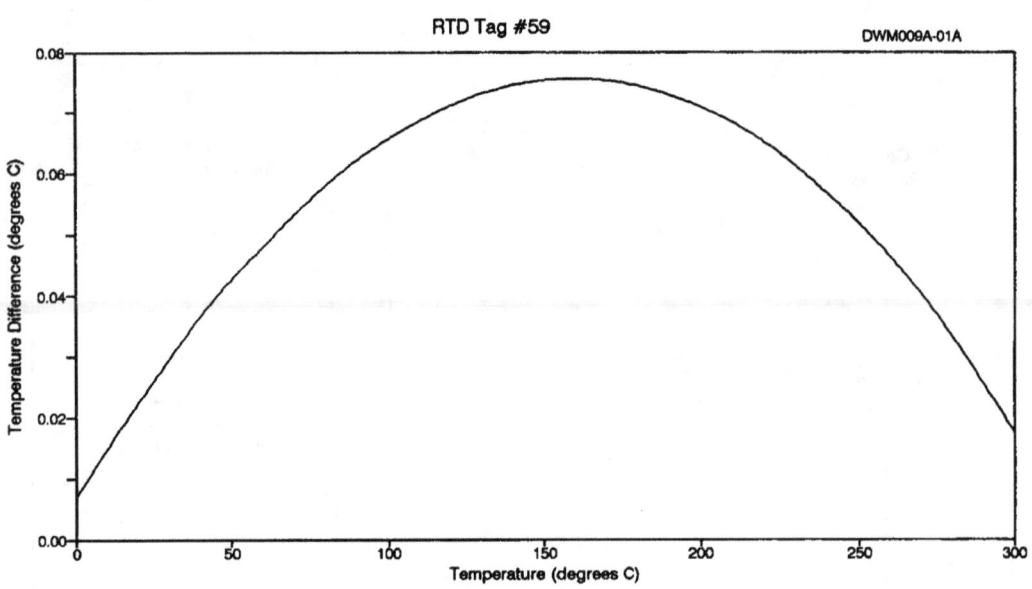

Figure 11-5. Examples of Calibration Difference Plots.

12. AGING TEST RESULTS FOR NUCLEAR GRADE RTDs

Laboratory tests were performed on 30 nuclear grade RTD elements aged at simulated reactor conditions. Thermal aging was emphasized because this is the most dominant stressor. Table 12.1 gives a listing of the RTDs and the aging categories in which they were included. The aging tests were performed over a 30 month period in the following order:

1. Thermal aging of all 30 elements for the first 18 months of the project.

2. Vibration aging of 10 RTDs and humidity aging of 11 RTDs for approximately two months each.

3. High temperature testing of 17 RTDs at 400°C for three days.

4. Thermal cycling of 19 RTDs performed over a two-week period.

The above test periods do not add up to 30 months due to set up time between the aging categories. The test results for each of the five aging categories are individually discussed below. These results correspond to the effects of normal operational aging as opposed to accelerated aging. At the conclusion of the normal aging tests, the RTDs were tested under severe aging conditions to determine their tolerance levels and failure modes. The results of these tests are given in Appendix A entitled "Destructive Testing".

12.1 Thermal Aging

Thermal aging tests were performed in two furnaces set at about 320°C. Pictures of the test setup are shown in Figures 12.1 and 12.2. The RTDs were installed in the furnaces and monitored continuously for gross performance problems during the aging process. Every one or two months, all RTDs were removed from the furnaces and calibrated. A full scope calibration was performed using measurements in an ice bath and an oil bath at 100, 200, and 300°C. The data were fit to the Callendar equation and the resistance versus temperature curves for each RTD was generated and stored. The results were compared with that of the initial calibration or the calibration from the previous month and the differences at 300°C were calculated. These differences are presented as the aging test results here and in all the sections that follow.

TABLE 12.1

Listing of Nuclear Grade RTDs Tested
for Normal Aging Effects

Tag	Thermal Aging	Vibration Aging	Humidity Aging	High Temperature	Thermal Cycling
3	◊	◊		◊	◊
4A	◊				
4C	◊ ♦				
5A	◊				
5C	◊				
7	◊		◊	◊	◊
8	◊				◊ ♦
9A	◊	◊		◊	◊
9C	◊	◊		◊	◊
11A	◊				◊
11C	◊		◊ ♦		
12A	◊		◊	◊	
12C	◊		◊	◊	◊
13A	◊	◊		◊	◊
13C	◊	◊		◊	◊
14	◊		◊		
15A	◊	◊		◊	◊
15C	◊	◊		◊	◊
16A	◊	◊		◊	◊
16C	◊	◊		◊	◊
17A	◊		◊	◊	◊
17C	◊		◊	◊	◊
18	◊ ♦				
19	◊		◊	◊	◊
20	◊				
21	◊	◊ ♦			
22A	◊		◊	◊	◊
22C	◊		◊	◊	◊
23	◊		◊		
24	◊				
Number Tested	30	10	11	17	19
Number Failed	2	1	1	0	1

◊ Denotes RTDs tested in each category.
♦ Denotes RTDs failed in each category.

AMS—DWG CAL008A

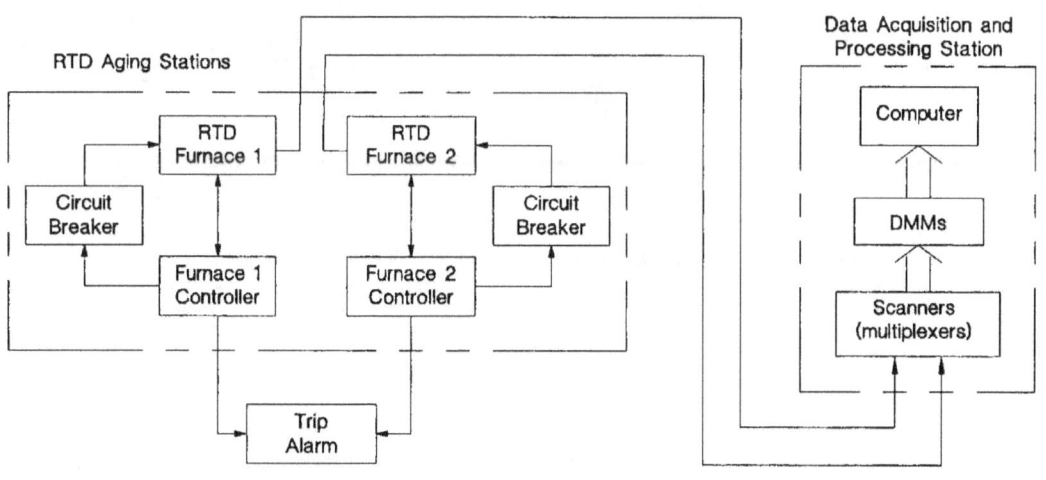

Figure 12-1. Thermal Aging Test Setup.

Figure 12-2. Close Up View of Thermal Aging Furnace.

The thermal aging tests were continued for 18 months during which each RTD was calibrated 8 to 12 times depending on when the RTD was placed in the project. The nuclear grade RTDs tested here included eight old RTDs from Phase I. At the beginning of the project, the old RTDs showed about twice as much drift as the new RTDs. This difference narrowed as the aging progressed and became insignificant after one year. Therefore, in presenting the results, we do not differentiate between the old RTDs and the new RTDs.

Two RTDs failed, six drifted as much as 0.6 to 3.0°C, and the rest settled into a ± 0.2°C drift band for the entire thermal aging period. The failures and most of the drifts occurred in the first six months of the project.

The drift behavior of the RTDs was as follows:

1.   Monotonic upward and downward drift as shown in Figure 12.3.

2.   Random drift in positive and negative directions within a finite band (Figure 12.4).

Random drift was the predominant behavior for most of the RTDs. As such, a single drift rate does not appropriately characterize the steady state behavior of the RTDs. Therefore, the drift data were characterized in terms of average values and drift ranges. Since there is both positive and negative drift, average values were calculated by two methods. One was to calculate the average of the absolute values and another was to calculate the average of the positive and negative values separately. Figure 12.5 shows the average drift of the unfailed RTDs after each calibration. The results are given in terms of the average of the positive and average of the negative drift values for all the RTDs excluding a few outliers. Based on the results shown in Figure 12.5, we can conclude that:

1.   The frequency and magnitude of drift of these RTDs are almost evenly distributed in the positive and negative directions.

2.   After an apparent burn-in period which lasted until the fifth calibration or approximately six months into the aging process, the RTDs stabilized in a drift band of less than ± 0.2°C. This point is supported by the data shown in Figure 12.6 and 12.7. Figure 12.6 gives the average of the absolute values of the individual drifts and Figure 12.7 gives the average of the positive and negative drifts for each RTD.

12.2 Vibration Aging

The vibration test setup is shown in Figure 12.8. It consists of a vibration table and a tube furnace in which the RTDs are vibrated while exposed to 320°C to provide the combined effects

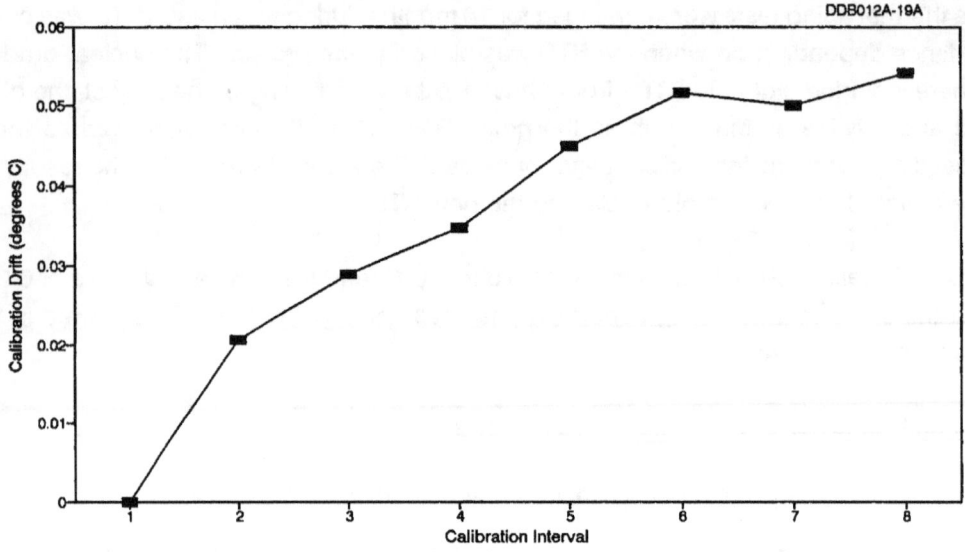

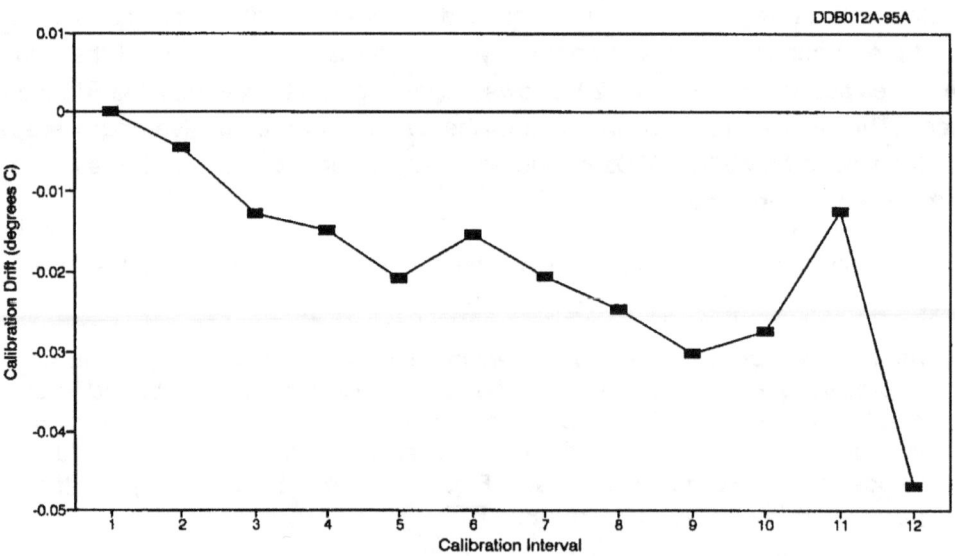

Figure 12-3. Monotonic Upward and Downward Drift Behavior.

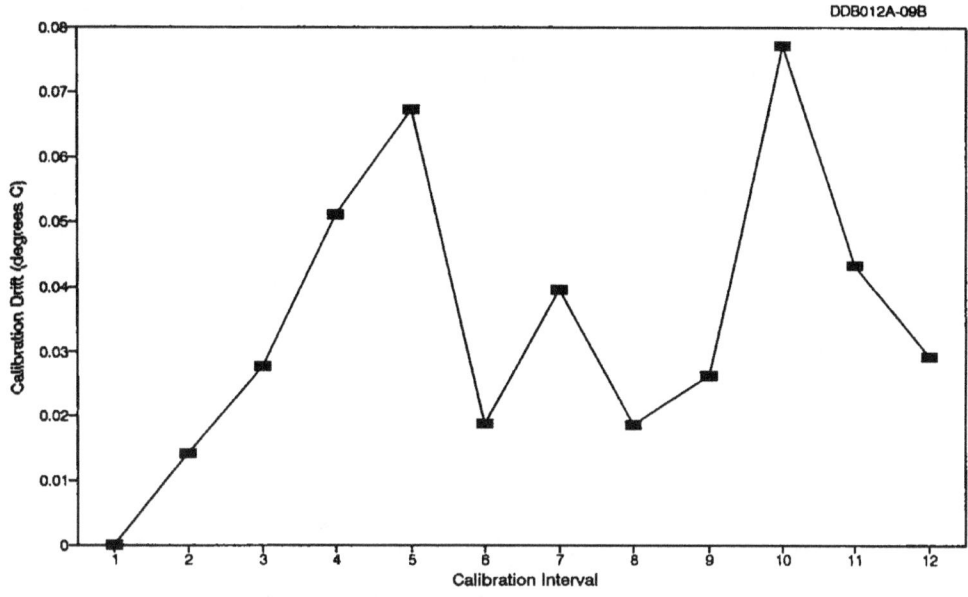

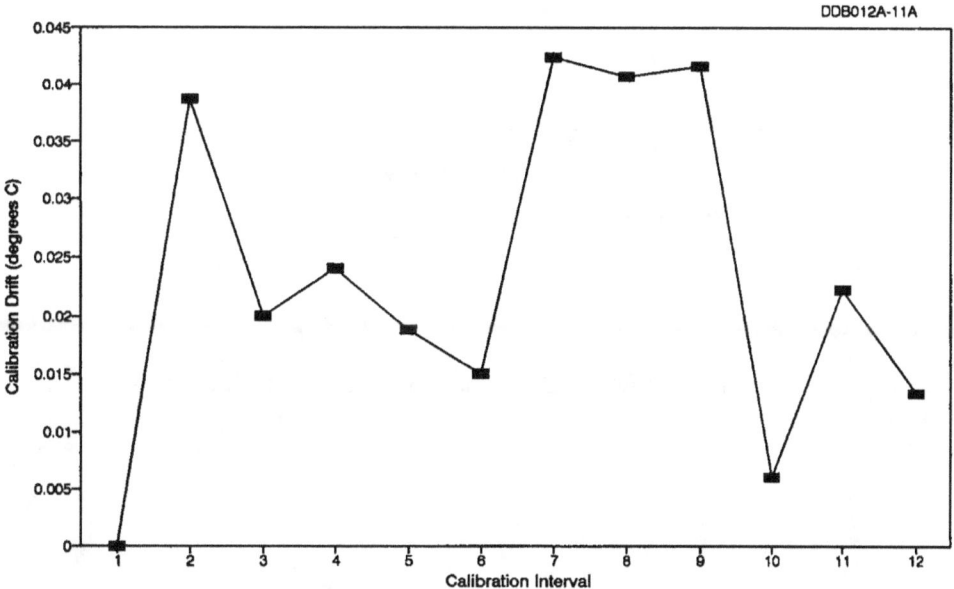

Figure 12-4. Two Examples of Random Drift.

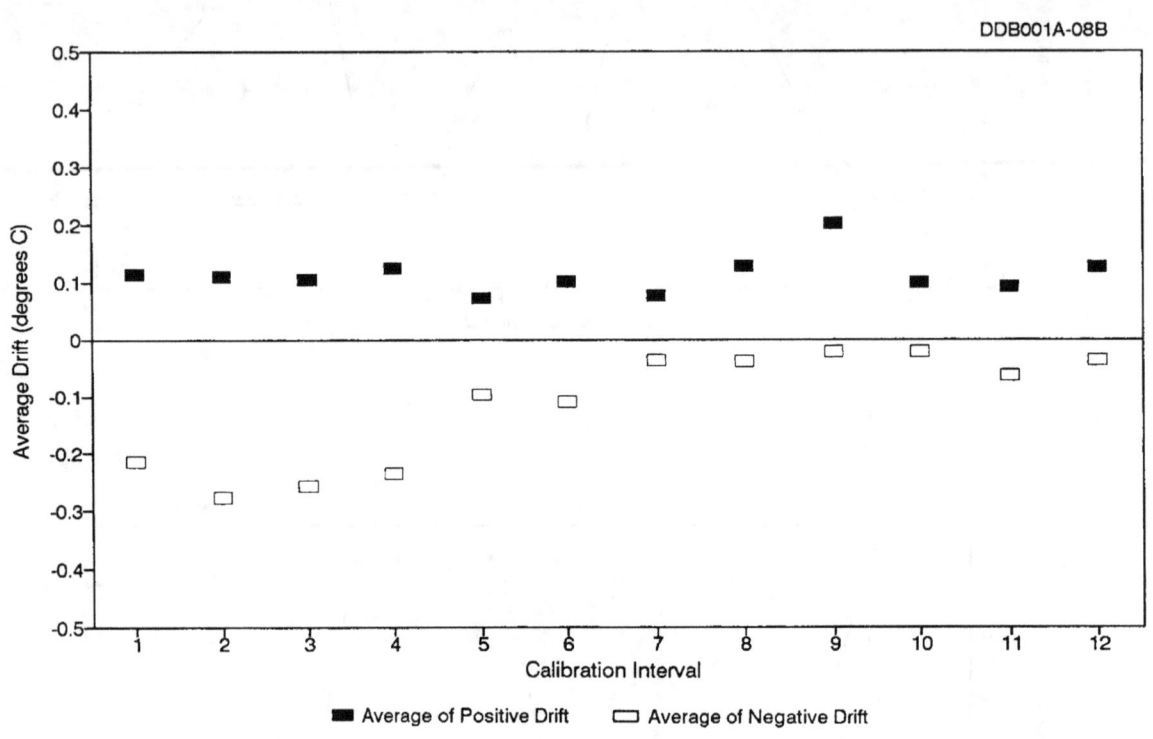

Figure 12-5. Thermal Aging Drift of the RTDs in the 18 Month Period.

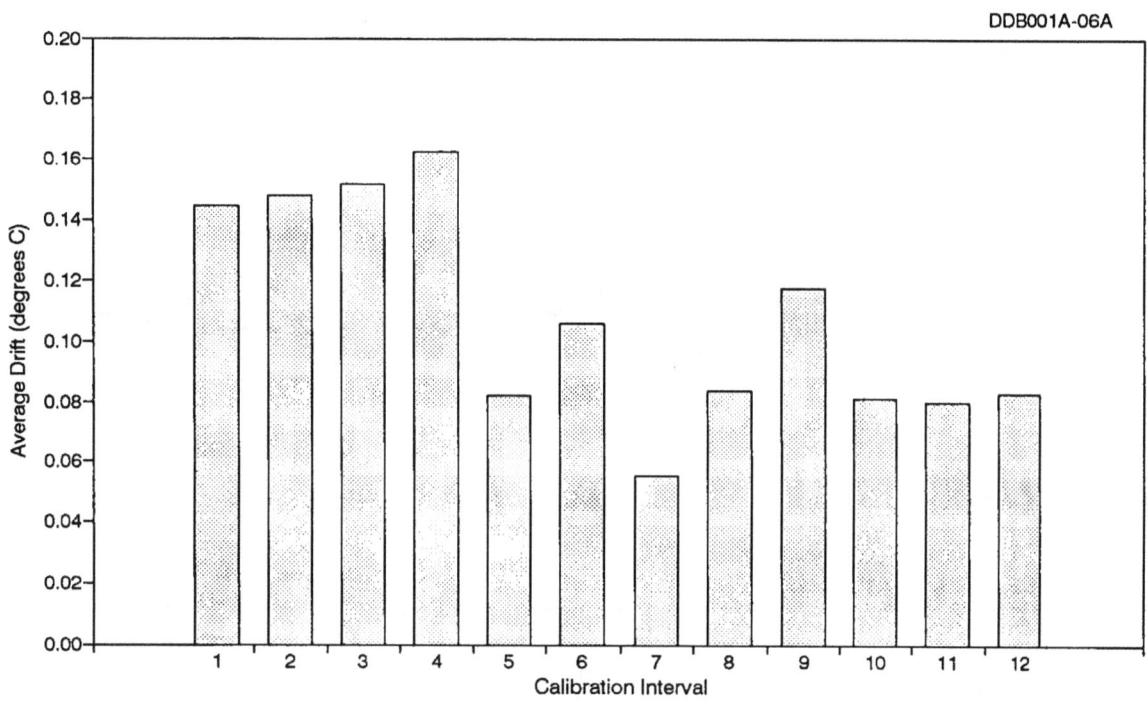

Figure 12-6. Average of the Absolute Values of Thermal Aging Drift.

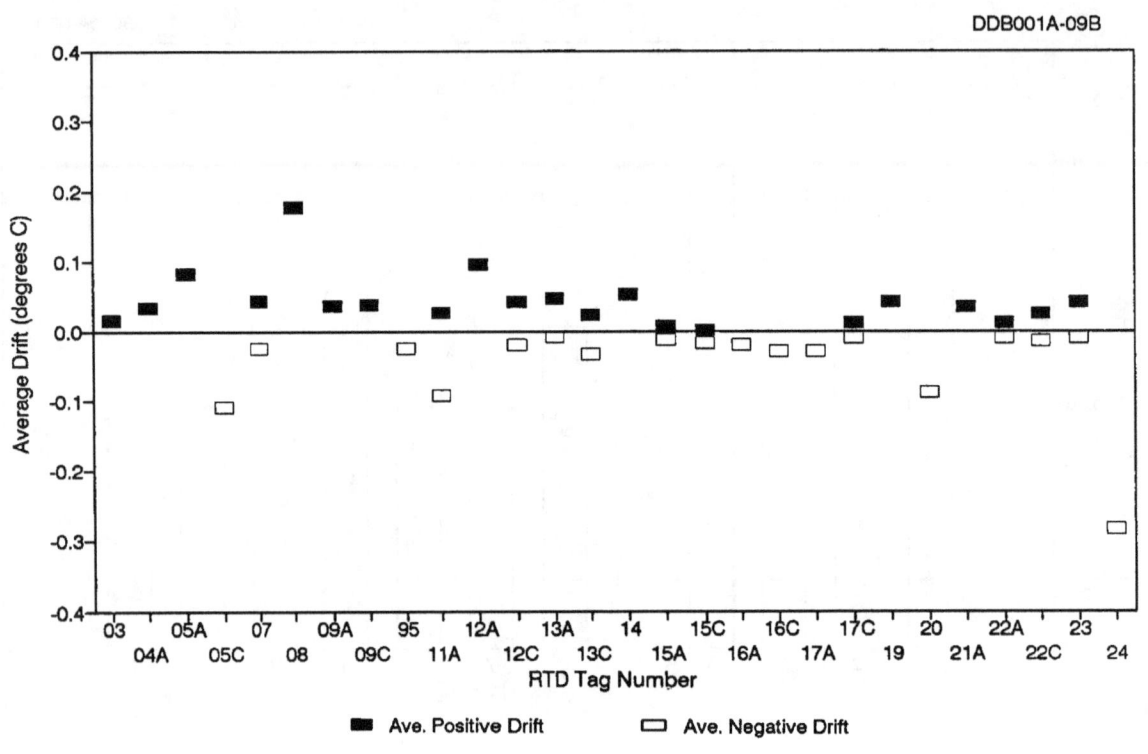

Figure 12-7. Average Thermal Aging Drift of
Individual RTDs.

Figure 12-8. Vibration Aging Setup.

of temperature and vibration. The vibration tests performed were intended to simulate the normal vibration induced by the flow of water in the primary coolant pipes and by nearby machinery as opposed to seismic vibration. The vibration parameters were set at an amplitude of approximately 3 millimeters and a frequency of about 20 Hz. This was selected based on our engineering judgment about the normal vibration levels to which RTDs are exposed in operating PWRs. The vibration tests were continued for 2 months on 10 RTDs. The RTDs were monitored continuously to identify any gross failures. One failure occurred and the remaining RTDs showed drifts of less than 0.06°C. Figure 12.9 shows the vibration drift of the RTDs in comparison with the drift for the entire thermal aging period.

In comparing the thermal and vibration aging results, one must note that the latter tests were performed over a two month period as opposed to 18 months for the former. A thorough analysis of the individual drift data from various aging tests performed in this project has indicated that the drift of these RTDs is not only random, but also interactive as opposed to cumulative. As such, the drift results shown in Figure 12.9 for two months of vibration aging shall not be viewed as 1/9 of the drift that would have resulted if the vibration tests were continued for 18 months. The purpose of showing the vibration test results in comparison with the thermal aging results is to demonstrate the relative effects of the two aging processes and to verify that long-term thermal aging results are not dominated by a few months of vibration aging.

12.3 Humidity Aging

The effect of humidity was studied in an environmental chamber as shown in Figure 12.10. Two series of tests were performed. The first series was performed at 65 percent humidity and a temperature of 50°C. Eleven nuclear grade RTDs were included in this experiment. These tests were continued for 30 days during which the RTDs were monitored every 30 minutes to identify any failure. This experiment did not result in measurable drifts.

In the next series of tests, the humidity level in the environmental chamber was increased to 90 percent at a temperature of about 60°C. The tests were continued for about two months. This produced one failure and less than 0.05°C drift except for one RTD which had a humidity drift of about 0.11°C. The results are given in Figure 12.11 in comparison with the thermal aging results. As discussed in Section 12.2, due to the random and nonadditive nature of the aging results, the humidity data in Figure 12.11 can not be extrapolated to estimate the drift of the RTDs for periods longer than two months.

12.4 High Temperature Tests

The high temperature tests were performed with 17 RTDs at 400°C for three days. This is the highest temperature for which most nuclear grade RTDs are rated. The tests did not produce any failures, and the average drift was less than 0.05°C except for RTD Number 7 and RTD Number 17 that had about 0.1 and 0.2°C drift respectively (Figure 12.12).

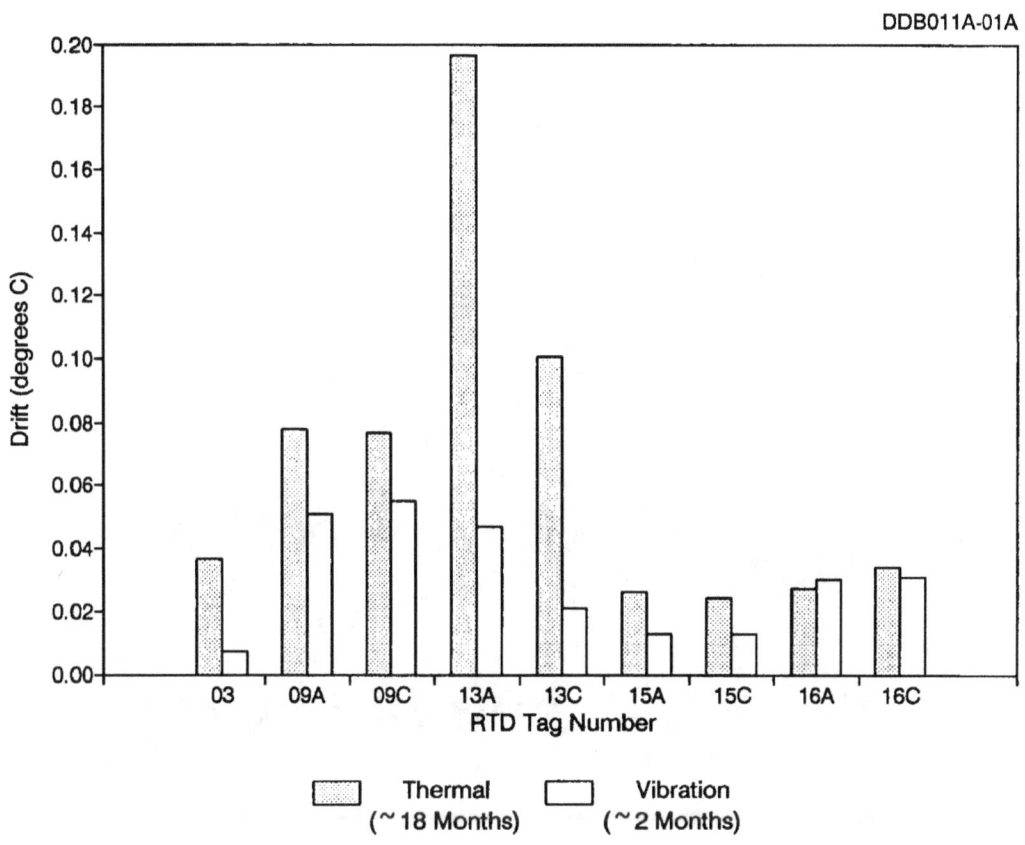

Figure 12-9.   Vibration Aging Drift Versus Thermal Aging Drift.

Figure 12-10. Humidity Aging Setup.

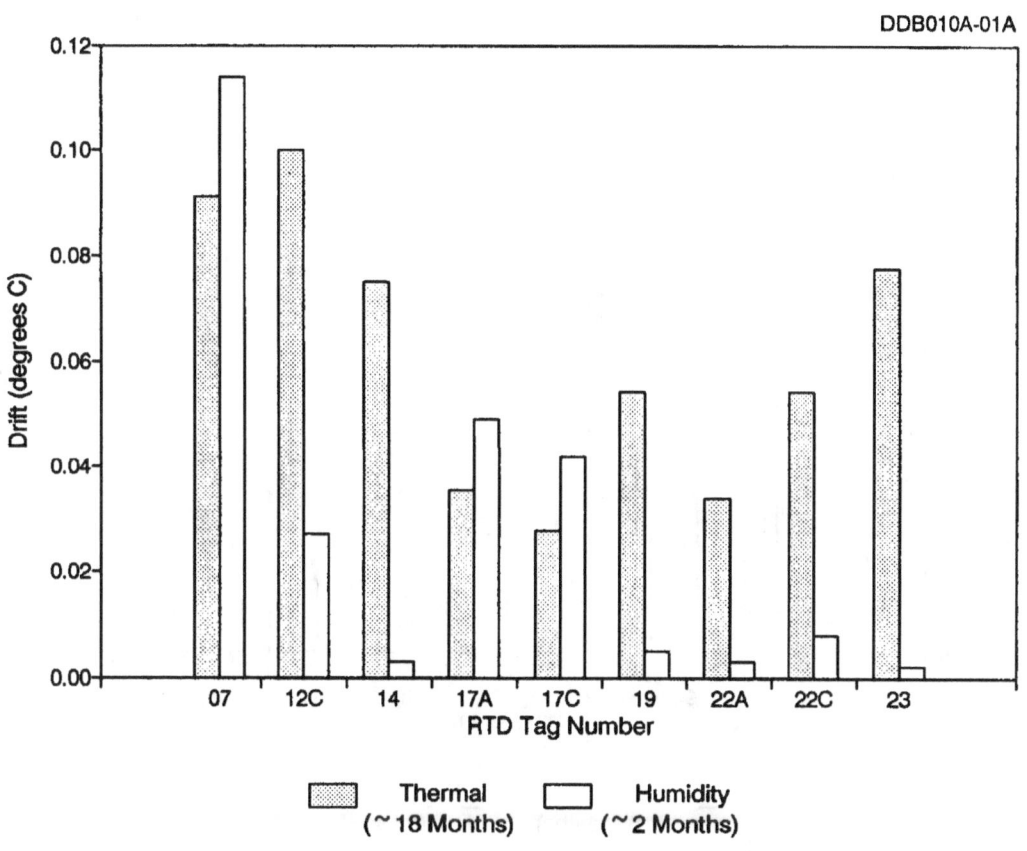

Figure 12-11.   Humidity Aging Drift Versus Thermal Aging Drift.

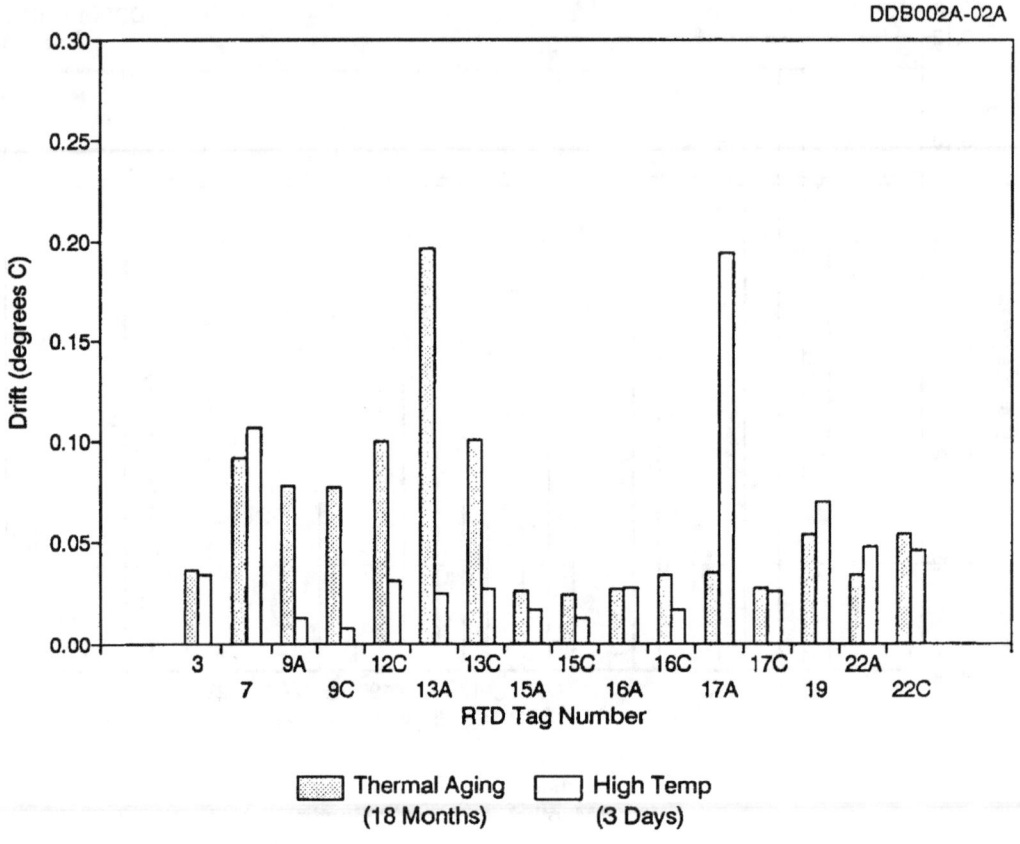

Figure 12-12.   High Temperature Drift Versus
                 Thermal Aging Drift.

Aging at 400°C for three days is equivalent to aging at 320°C for about two years using the "10°C" rule. If this rule applies to RTDs, then we have demonstrated not only by actual aging for about two years, but also by accelerated aging (equivalent to about two years) that the thermal aging drift of RTDs is contained within a ± 0.2°C band due to the effects of thermal aging at PWR operating temperatures. The "10°C" rule has been derived from the Arrhenius theory and is used as follows: the life of a component at a reference temperature approximately doubles for every 10°C increment in temperatures above the reference temperature. Note that the results of the high temperature tests as well as the vibration and humidity tests discussed before are probably influenced by the fact that the RTDs have undergone a "burn-in" experience during the thermal aging process with which they began their project life.

12.5 Thermal Cycling

Thermal cycling tests were performed by plunging the RTDs in an oil bath. The tests involved 19 RTDs which were cycled between room temperature and 300°C every 30 minutes. The tests were continued for 100 cycles. This resulted in one failure and calibration drifts of less than 0.2°C (Figure 12.13). Note that the cycling conditions used here were more severe than those experienced by RTDs during normal plant operation in which RTDs are cycled only at heat up and shutdown periods and during plant trips.

12.6. Summary of Aging Test Results

The aging test results are summarized in Figure 12.14. The results shown are the average drifts from the end of one test to the end of the next test. The application of the five aging processes on 10 to 30 RTD elements tested here produced five failures (17%) and six cases (20%) of drift in the range of 0.6 to 3.0°C. The cause of the failure of each RTD is shown in Table 12.2. The remaining 19 RTDs (63%) performed well, with their normal aging drifts contained within a ± 0.2°C band. This conclusion excludes the drift that occurred in the first few months of thermal aging while the RTDs were presumably undergoing a "burn-in" experience. Note that the results of vibration, humidity, and other tests that were performed after thermal aging are probably affected by any stability that could have resulted from the "burn-in" period in the furnaces.

The aging of the RTDs did not result in a monotonic drift from which a reliable drift rate could be obtained. It was demonstrated that none of the normal aging conditions alone can generally produce more than an average of 0.2°C drift and combining the aging effects will not increase the drift significantly beyond the largest drift from individual effects.

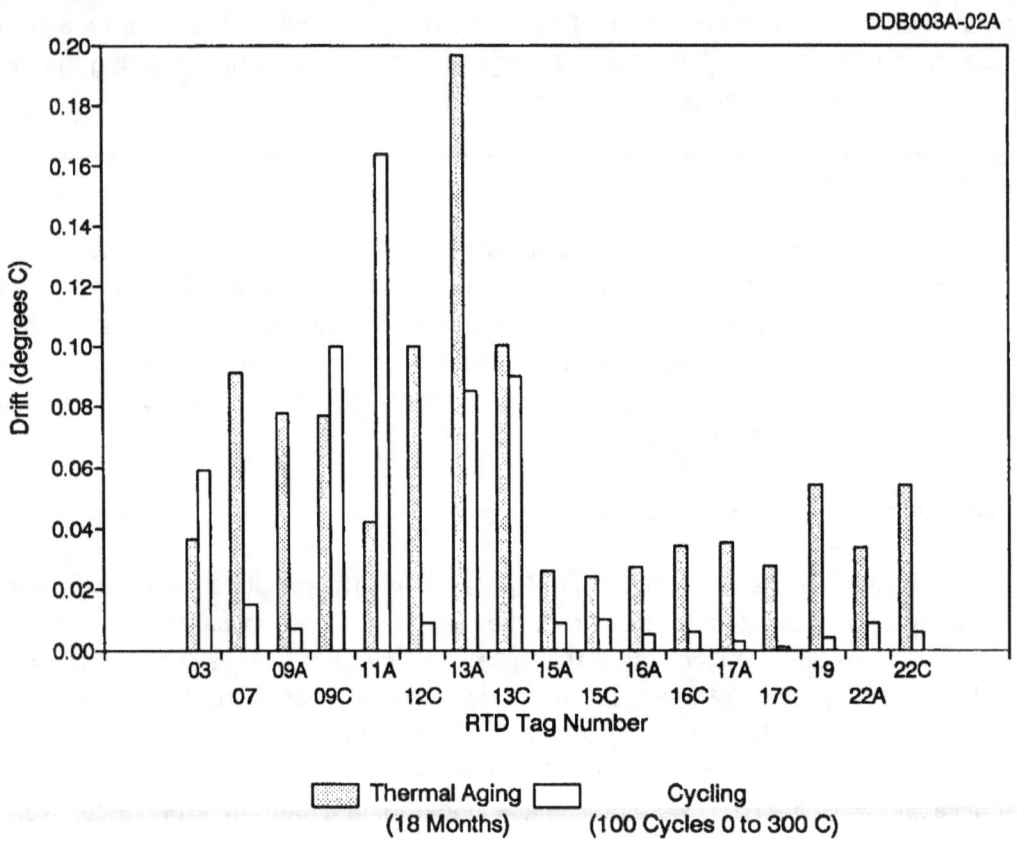

Figure 12-13.   Temperature Cycling Drift Versus
                 Thermal Aging Drift.

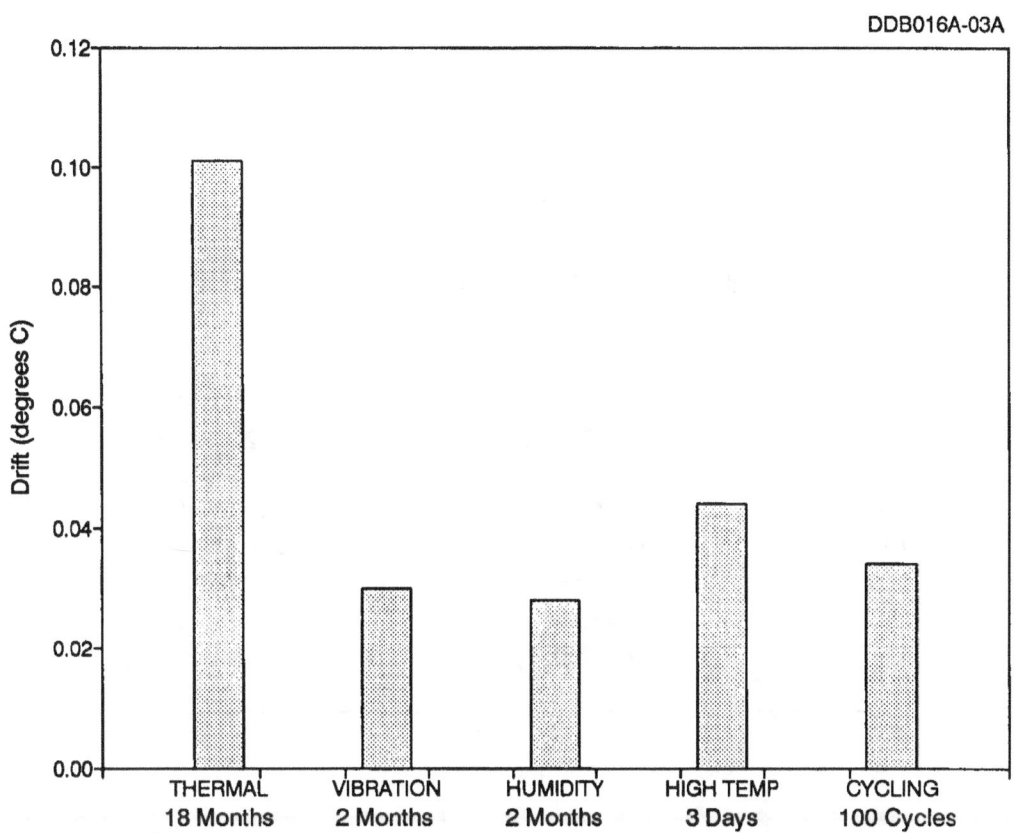

Figure 12-14.   Summary of Aging Test Results.

TABLE 12.2

Listing of RTDs Failed in Normal Aging

Tag	Aging Category	Reason for Failure
4C	Thermal	> 5°C Shift
18	Thermal	> 5°C Shift
11C	Humidity	> 5°C Shift
21C	Vibration	> 5°C Shift
8	Thermal Cycling	Open circuit

13. AGING TESTS OF COMMERCIAL GRADE RTDs

A listing of the commercial grade RTDs and the aging tests performed on these RTDs is given in Table 13.1. These RTDs were exposed to approximately the same aging conditions discussed in Section 12 for the nuclear grade RTDs. One failure (Tag No. 28) occurred two months into the thermal aging process. The cause of this failure was an open circuit. The only other failure was that of Tag No. 25 which occurred in humidity aging. This RTD shifted by 5°C. Figures 13.1 and 13.2 give the thermal aging drift of the commercial grade RTDs in comparison with thermal cycling and high temperature aging. Results are not shown for vibration and humidity aging as there were very few commercial grade RTDs included in these processes.

The aging test results for the nuclear and commercial grade RTDs are compared in Figure 13.3. It is apparent that the performance of nuclear grade RTDs is better than that of commercial grade RTDs by a factor of nearly two. The results are shown for three aging episodes. The first one is that of normal aging discussed here and in Section 12. The second and third are for aging to extremes of normal conditions and aging to study failure modes as discussed in Appendix A. The results shown in Figure 13.3 represent the averages of the drifts from all aging categories for all the RTDs involved in each category. Note that the drift bands for the combined aging results shown in Figure 13.3 for the commercial grade RTDs are less than the drift bands for single mode aging shown in Figures 13.1 and 13.2. This is because the results shown in Figure 13.3 are the averages for a number of RTDs, some of which did not show much drift. Furthermore, the individual aging effects may work against one another and thereby cause a smaller overall drift than the drift for individual effects.

TABLE 13.1

Commercial Grade RTDs Included in the Normal Aging Tests

Tag	Thermal Aging	Vibration Aging	Humidity Aging	High Temperature	Thermal Cycling
25	◊		◊ ♦		
26	◊		◊	◊	◊
27	◊	◊		◊	◊
28	◊ ♦				
35	◊				
36	◊			◊	◊
37	◊				◊
38					
42	◊		◊	◊	◊
43	◊		◊		
44	◊			◊	◊
51	◊	◊		◊	◊
52	◊			◊	◊
53	◊		◊	◊	◊
54	◊			◊	◊
95	◊				
48	◊				
Number Tested	16	2	5	9	10
Number Failed	1	0	1	0	0

◊ *Denotes RTDs tested in each category.*
♦ *Denotes RTDs failed in each category.*

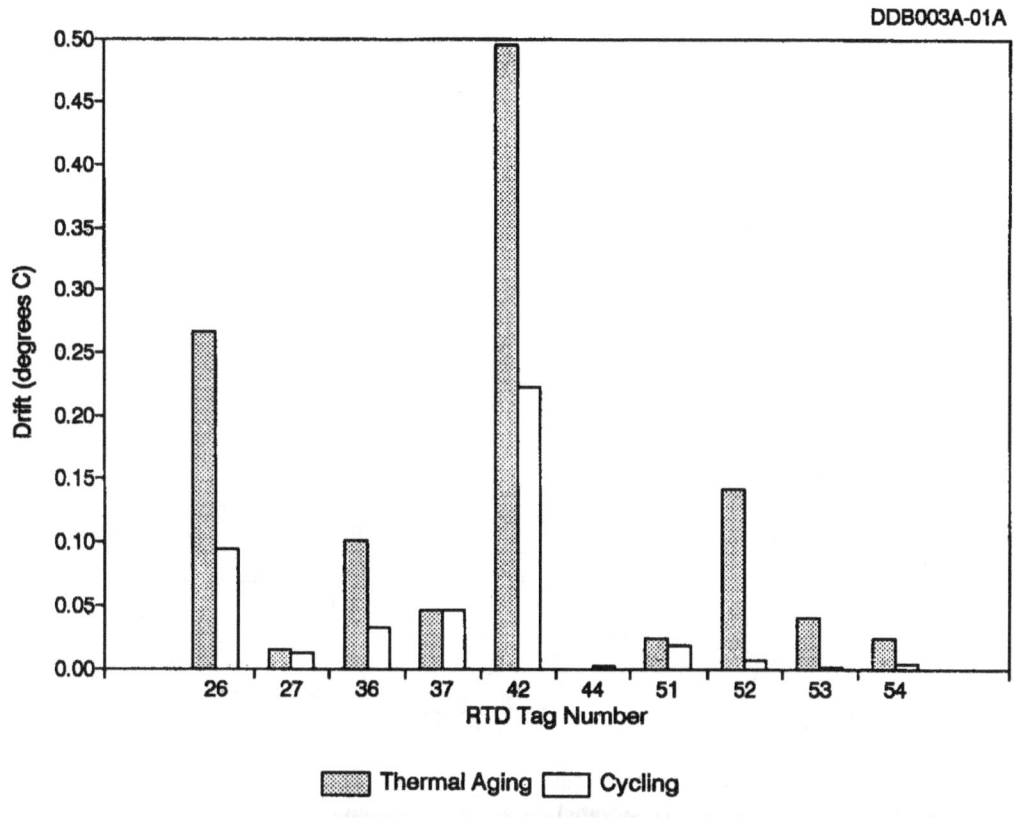

Figure 13-1.   Thermal Cycling Drift of Commercial Grade RTDs
Compared With Their Thermal Aging Drift.

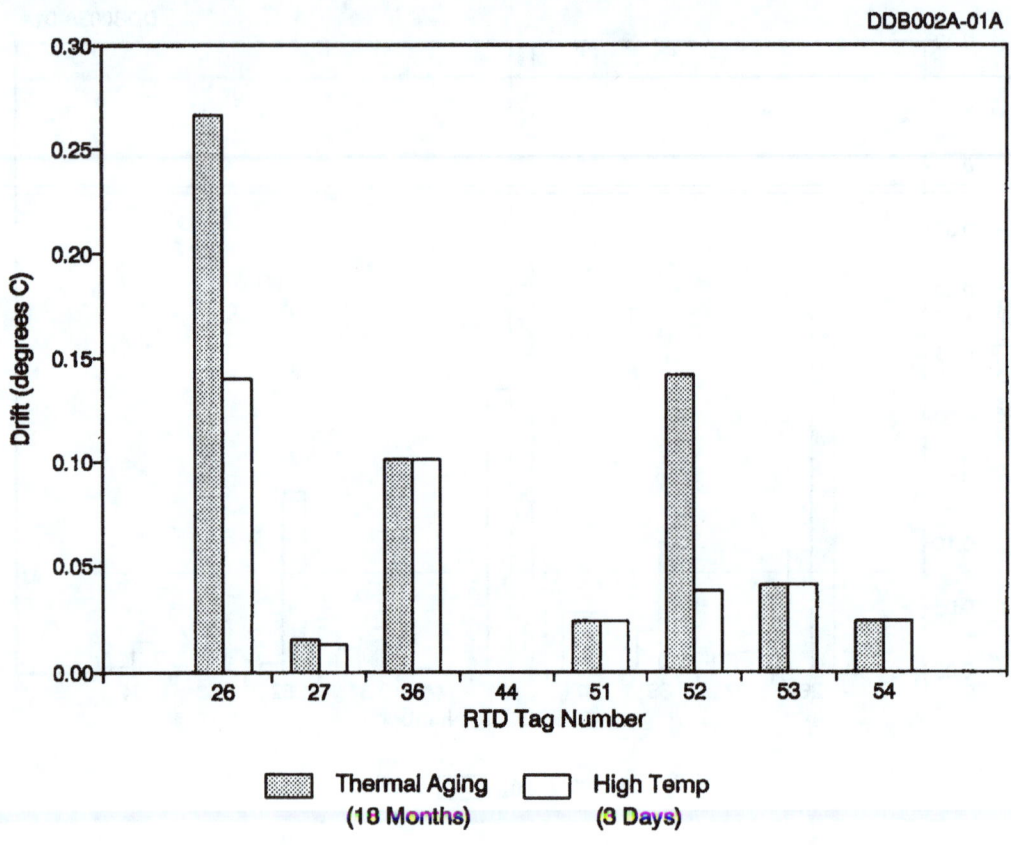

Figure 13-2.   High Temperature Drift of Commercial Grade
                RTDs Compared With Thermal Aging Drift.

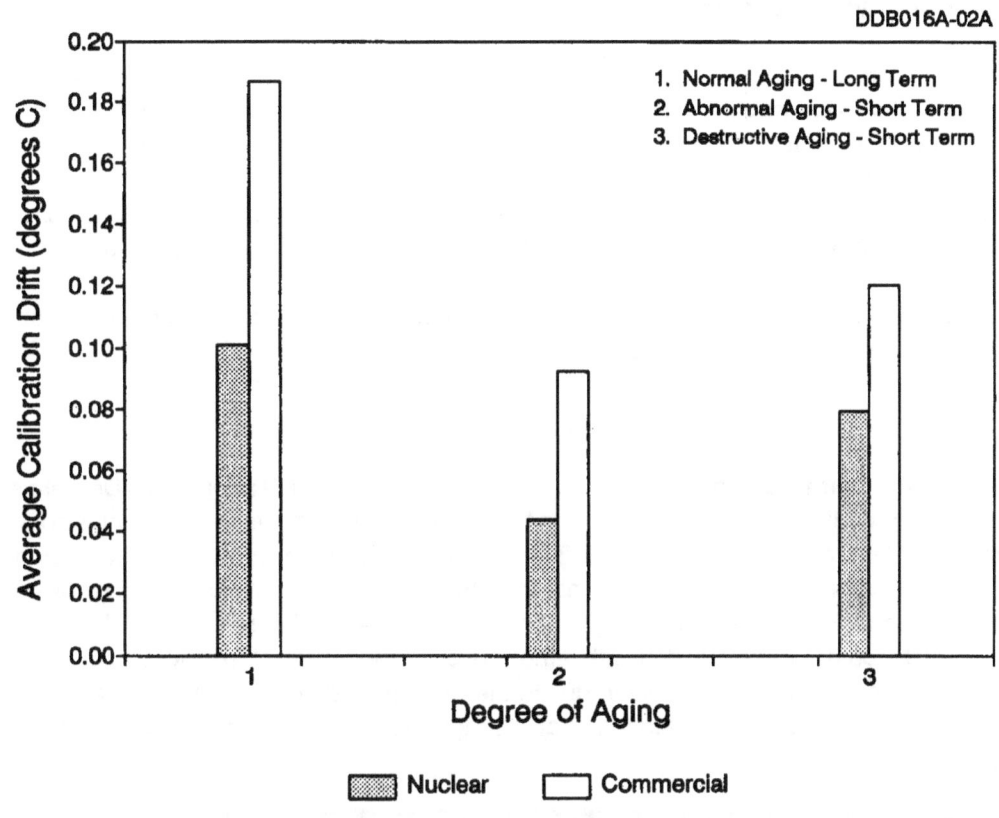

Figure 13-3.   Comparison of Aging Drift of Nuclear and
                Commercial Grade RTDs.

14. SHELF-LIFE DRIFT

Nuclear plant RTDs sometimes remain in storage or in the plant for a few months or a few years before they are used. This raises a question about any drift that may occur while the RTDs are not in active service. This question is addressed here.

14.1 Drift of Naturally Stored RTDs

Table 14.1 provides storage drift data for 24 nuclear grade RTDs. These RTDs were in warehouses of various nuclear power plants for periods ranging from one to five years. The exact storage age of each RTD is not known. The results represent the differences between the initial factory calibration and the calibration performed at AMS as each RTD arrived. The average drifts shown in Table 14.1 are less than about 0.1°C which is approximately half as much drift as we identified for thermal aging at plant operating conditions for a two-year period.

The results given in Table 14.1 are the differences between the pre- and post- shelf-life calibration of the RTDs at 100, 200, and 300°C.

14.2 Laboratory Test Results

Following the thermal aging tests discussed in Section 12, 21 nuclear grade RTDs were stored for four months during which they were monitored and calibrated every two months to identify shelf-life drift. The results are shown in Figure 14.1 in comparison with the thermal aging drifts discussed in Section 12. It is apparent that the storage drift is significant. The results for the commercial grade RTDs are similar (Figure 14.2). Note that in Figures 14.1 and 14.2, we are comparing 18 months of thermal drift with 4 months of storage drift. Normally, a comparison like this would not be reasonable. However, due to the random and non-additive nature of RTD drift, we made the comparison to illustrate the relative magnitudes of the drifts for thermal and storage aging.

In another test for storage drift, six RTDs were repeatedly calibrated one day apart and then one week apart. The results are shown in Table 14.2 in terms of the calibration drifts identified over a 24-hour period and over a seven day period. The latter produced about six times as much drift as that of 24-hours. Of course, the storage drift does not continue to climb. If it did, the four months of storage drift data shown in Figure 14.1 would have had much larger values. Based on the data shown here and other data produced in the project, it has been concluded that the storage drift of RTDs begins with a high rate and settles at a point of about 0.1°C.

The storage drift tests reported here were conducted with the RTDs on the bench at room temperature, pressure, and humidity.

TABLE 14.1

Shelf-Life Drift Measured for Nuclear
Grade RTDs at Three Temperatures

Tag	Calibration Drift (°C)		
	100°C	200°C	300°C
62	-0.12	-0.17	-0.13
64	0.04	0.08	0.11
63	-0.03	-0.24	-0.68
65	-0.10	-0.12	-0.05
77A	0.08	0.08	0.02
77C	0.15	0.15	0.01
78A	0.06	0.07	0.03
78C	0.06	0.07	0.02
79A	0.06	0.06	-0.00
79C	0.07	0.06	-0.00
80A	0.07	0.09	0.07
80C	0.06	0.09	0.09
81A	0.04	0.04	0.00
81C	0.04	0.03	-0.02
82A	0.10	0.10	-0.01
82C	0.12	0.10	-0.05
83A	0.16	0.15	-0.02
83C	0.15	0.16	0.02
84A	0.14	0.18	0.13
84C	0.13	0.17	0.14
11A	-0.11	-0.11	-0.04
11C	-0.13	-0.13	-0.04
12A	-0.10	-0.08	0.04
12C	-0.14	-0.13	0.01
Average	0.09	0.11	0.07

Above RTDs have been in storage for about 1 to 5 years.

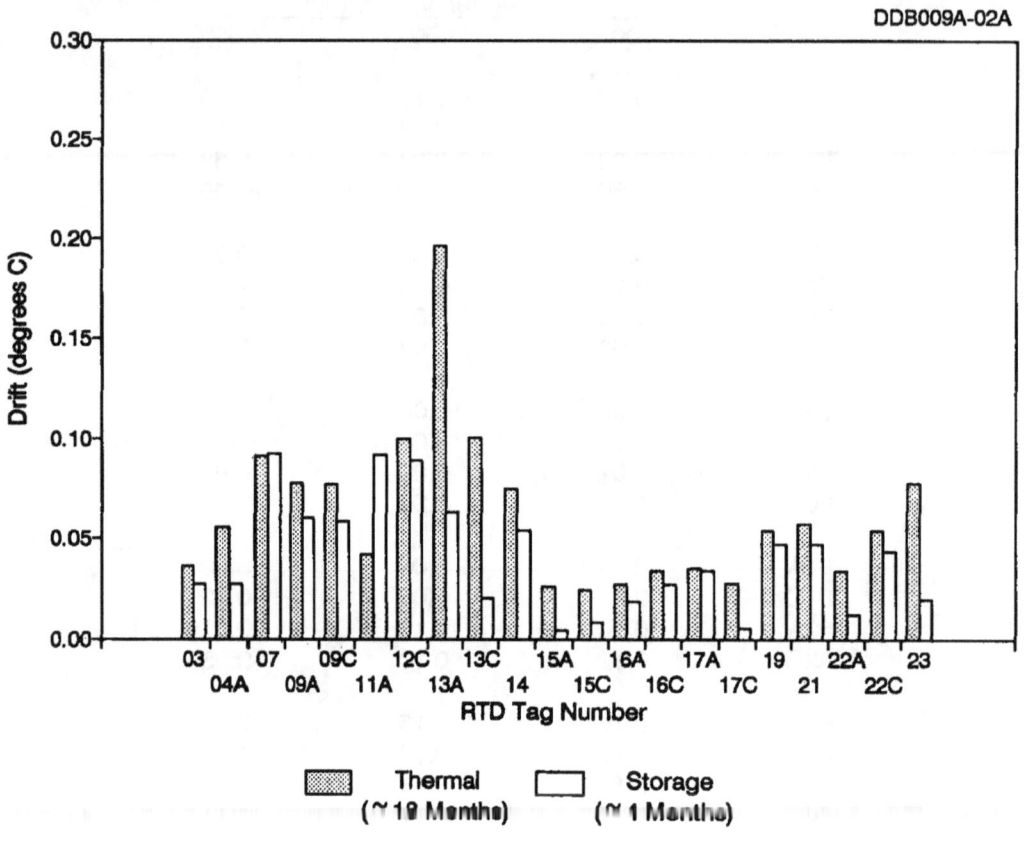

Figure 14-1.   Storage Drift of Nuclear Grade RTDs Compared
With the Drift During the Entire Thermal Aging
Period.

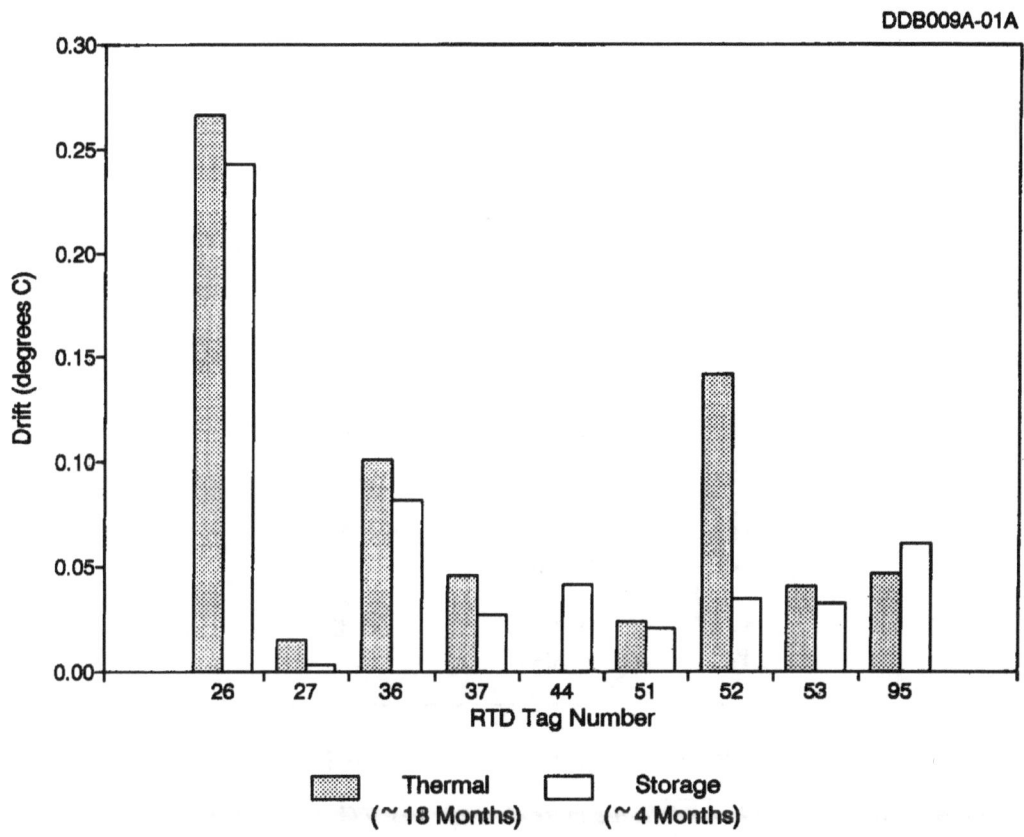

Figure 14-2.   Storage Drift of Commercial Grade RTDs Compared
With Their Thermal Aging Drift.

TABLE 14.2

Shelf-Life Drift as Determined by One Day and One Week Repeatability of Nuclear Grade RTDs

Tag	Drift (°C) One Day	One Week
15A	0.009	0.06
15C	0.011	0.06
16A	0.004	0.06
16C	0.005	0.05
17A	0.008	0.02
17C	0.009	0.03
Average	0.008	0.05

Above drifts are those observed at 300°C.

15. TESTING OF NATURALLY AGED RTDs

Eight nuclear grade RTDs removed from various plants were calibrated and the results were compared with that of the original factory calibration. These RTDs were used in normally operating PWRs for two to five years. The exact age of each RTD is not known. The results are shown in Table 15.1 in terms of the differences between the new calibration and the original factory calibration at four temperatures (0, 100, 200, and 300°C). These results correspond to the drift of the RTDs due to natural aging in the plant and any drift that might have occurred from when each RTD was removed from the plant to when it was calibrated at AMS. Note that the average drift at 300°C is about 0.2°C which is consistent with the laboratory aging results discussed in Section 12.

Another opportunity for determining the drift of naturally aged RTDs was provided in two series of cross calibrations performed in a PWR with sixteen primary coolant RTDs. In the first series, one RTD was found with 1.4°C deviation and four were flagged as having more than 0.3°C deviations from the average of all RTDs. These five were replaced and the cross calibrations repeated at three temperature plateaus. The results were used to identify the drift of the remaining eleven RTDs. The results are shown in Table 15.2 in terms of the drifts identified at three temperature plateaus where cross calibrations were performed. These results represent the drift of the RTDs for five years of natural aging in an operating nuclear power plant. The average drift for the eleven RTDs is 0.13°C. Again, this is reasonably consistent with the results of the laboratory aging tests.

TABLE 15.1

Drift of Naturally Aged RTDs
Identified By Laboratory Calibration

Tag	Drift (°C)			
	0°C	100°C	200°C	300°C
57	0.004	0.14	0.17	0.07
85	0.015	0.02	0.00	0.04
86	0.007	0.01	0.36	1.13
75A	-0.024	0.02	0.04	0.02
76C	0.023	0.09	0.10	0.03
36	0.045	-0.07	-0.09	-0.01
37	0.060	0.05	0.06	0.09
38	0.070	0.03	0.05	0.12
Average	0.031	0.05	0.11	0.19

Above RTDs have been used in various plants for about 2 to 5 years.

TABLE 15.2

Drift of Naturally Aged
RTDs Identified by Cross Calibration

| | Drift (°C) | | |
Tag	280°C	220°C	160°C
1	-0.09	-0.07	-0.11
2	0.08	0.09	0.04
3	0.05	0.06	0.10
4	0.07	0.01	0.09
5	-0.38	-0.37	-0.23
6	0.01	0.05	0.00
7	0.02	0.00	-0.02
8	0.08	-0.07	-0.15
9	-0.13	-0.12	-0.24
10	0.26	0.22	0.26
11	-0.21	-0.24	-0.20
Average	0.13	0.12	0.13

The RTDs shown here have been used in the plant for about 5 years.

16. ACCURACY OF RESISTANCE VERSUS TEMPERATURE TABLES

The initial accuracy of an RTD is established by combining the uncertainties involved in its laboratory calibration. Three groups of uncertainties must be included:

- Calibration equipment uncertainties.
- Inherent RTD errors such as hysteresis, repeatability, and self heating.
- Interpolation and fitting errors.

Table 16.1 gives the ranges of the uncertainties and the discussions that follow describe how they are identified and combined.

16.1 Calibration Equipment Uncertainties

The calibration of an RTD in the range of 0 to 300°C involves comparing its resistance with that of an SPRT in an ice bath and an oil bath at several temperatures. The following uncertainties are involved in this process.

- Ice Bath Stability and Uniformity. The maximum temperature difference between the SPRT and the RTD being calibrated depends on the bath stability and uniformity. The ice bath stability and uniformity was measured with two SPRTs, one at the center of the bath and another about one inch from the center. This placement corresponds to the normal locations of the SPRT and the RTDs being calibrated. The measurements showed that the average temperature difference between the two locations is 0.001°C and the ice bath stability at each location is 0.001°C. The stability was determined by calculating the standard deviations of the ice bath fluctuations as indicated by the two SPRTs. The uniformity plus one standard deviation for each SPRT was 0.003°C. This represents a conservative estimate for the maximum difference that may exist between any two RTDs in the ice bath.

- Oil Bath Stability and Uniformity. In Section 11, we showed how the oil bath stability and uniformity is improved using a copper block in which the RTDs are calibrated. Figure 16.1 shows the stability and uniformity of the oil bath at 300°C measured with two SPRTs, one in the center of the copper block and one in the perimeter. The average difference is 0.003°C and the standard deviation is also 0.003°C for each SPRT. The average plus one standard deviation for each SPRT is 0.009°C. This is the typical difference that can exist between the SPRT at the center of the copper block in the oil bath and any location in the block perimeter. The bath stability and uniformity results obtained by measurements using two SPRTs are shown in Table 16.2 for an ice bath at 0°C and an oil bath at 100, 200, and 300°C.

TABLE 16.1

RTD Calibration Uncertainties for the
Temperature Range of 0 to 300°C

	Sources of Uncertainty	Range (°C)	Remarks
1.	Bath Stability and Uniformity	0.009-0.02	1
2.	SPRT Accuracy and Drift/Year	0.005-0.01	2
3.	Measurement Equipment for SPRT Accuracy and Drift	0.005-0.01	3
4.	Measurement Equipment for RTD Accuracy and Drift	0.005-0.01	3
5.	Hysteresis	0.010-0.30	4
6.	Repeatability	0.010-0.10	4
7.	Self Heating	0.001-0.01	5
8.	Fitting Errors	0.005-0.01	6

Remarks

1. Based on measurements made at 300°C in a copper block inside a well-stirred oil bath.

2. Based on three years of historical calibration data on two SPRTs and information available in the literature and from NIST.

3. Based on manufacturer's data for high precision digital multimeters.

4. Based on measurements made of two groups of nuclear grade RTDs.

5. Based on differences in self heating indices of five RTDs in an oil bath and in water flowing at 1 meter/sec.

6. Based on measurements and theoretical analysis of typical calibration data.

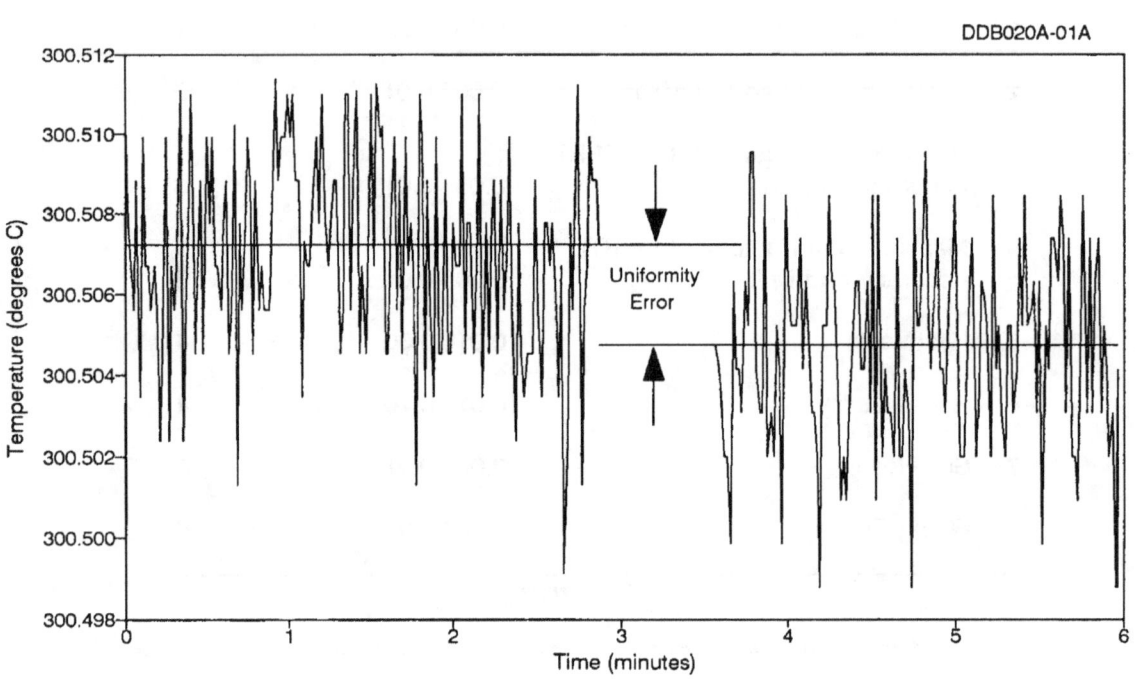

Figure 16-1. Stability and Uniformity Data for 300°C Oil Bath.

TABLE 16.2

Results of Bath Stability and Uniformity Measurements

Temperature (°C)	Stability (°C)		Uniformity (°C)	Total Uncertainty (°C)
	σ_1	σ_2		
0	0.001	0.001	0.001	0.003
100	0.002	0.002	0.002	• 0.006
200	0.002	0.002	0.003	0.007
300	0.003	0.003	0.003	0.009

σ_1: *Standard deviation of bath fluctuations for 25 samples obtained with SPRT-1 in the center of a copper block in the oil bath. An equalizing block was not used in determining the ice bath stability and uniformity.*

σ_2: *Standard deviation of bath fluctuations for 25 samples obtained with SPRT-2 in the perimeter of a copper block in the oil bath.*

Note that both the bath stability and uniformity are dominated by that of the oil bath at 300°C. Therefore, the table of uncertainties (Table 16.1) includes only the oil bath stability and uniformity. Based on the data at 300°C, the lower end of the bath stability error is 0.009°C and the upper end is 0.02°C obtained by adding three standard deviations for each of the two SPRTs to the bath uniformity error at 300°C.

We also measured the oil bath stability and uniformity without the copper block. The combined stability and uniformity error was 3 to 5 times larger than the errors when the copper block is used. Although the use of a copper block is important for precision RTD calibration, in cases where accuracies of greater than 0.5°C are acceptable, a copper block does not have to be used especially since it limits the number of RTDs that can be tested at a time and also because it increases the self heating error.

• SPRT Accuracy and Drift. An SPRT that has been calibrated at NIST has an initial accuracy of 0.001 to 0.005°C. The drift of an SPRT is 0.005 to 0.010°C per year depending on the SPRT.

Sometimes a secondary standard is used in performing the calibration. In this case, the accuracy and drift of the secondary standard must be used rather than that of the SPRT. The accuracy and drift of secondary standards is typically about 0.05 and 0.10°C respectively.

• Measurement Equipment Error and Drift. The resistance of an RTD is measured with a resistance bridge or a precision digital multimeter. Multimeters are needed if the calibration is to be done automatically with the aid of a computer. The accuracy of precision equipment for resistance measurement corresponds to 0.005 to 0.01°C in temperature, including short term drift.

16.2 Hysteresis Error

Thermal hysteresis is caused by differences in thermal expansion coefficients between the platinum element and the structure in contact with the element. Errors due to hysteresis are dependant on the prior temperature history experienced by the RTD. More specifically, the resistance at a calibration point will be different depending on whether the RTD was exposed to higher or lower temperatures prior to reaching the calibration point. Hysteresis errors were measured in two groups of RTDs by taking calibration points when the bath temperature was increasing and comparing the results with those obtained when the bath temperature was decreasing. The calibrations were performed over the 0 to 300°C range. Twelve calibration points were taken with increasing temperatures and twelve with decreasing temperatures. The results are shown in Table 16.3 for two groups of nuclear grade RTDs designated as new RTDs and old RTDs. The average is about 0.01°C for the new RTDs and 0.3°C for the old RTDs. The old RTDs are those that remained from the Phase I project. It is not known why the two groups of RTDs have two distinct values for hysteresis error. The 0.01 to 0.3°C range found for

TABLE 16.3

Hysteresis Errors in
New and Old RTDs

Tag	Hysteresis Error (°C)
New RTDs	
2	0.002
15A	0.015
15C	0.017
16A	0.022
16C	0.017
17A	0.005
17C	0.015
95	0.004
Old RTDs	
3	0.25
4A	0.34
4C	0.32
5A	0.35
5C	0.35
7	0.08
8A	0.27
9A	0.32

hysteresis error is consistent with information available in literature and from manufacturers for hysteresis of industrial RTDs.

Table 16.4 shows hysteresis repeatability of the old RTDs. The errors were identified on three occasions as shown by the date of each test. The table also shows the average drift of the RTDs due to thermal aging. It is apparent that the hysteresis errors are independent of the RTD drift characteristics.

Hysteresis errors can be minimized by taking calibration points with increasing and decreasing temperatures and averaging the results even though this approach was not used in the calibrations performed in this project.

Figure 16.2 demonstrates hysteresis effects in the range of 0 to 300°C for two RTDs. This data was obtained by cycling the RTDs between 0 and 300°C. Note that the RTDs do not return to the same end points because of potential hysteresis and repeatability effects.

16.3 Repeatability Error

Repeatability error depends on the stability of the RTD and repeatability of the calibration process. For stable RTDs, a precise calibration yields repeatabilities of better than 0.01°C. Table 16.5 shows repeatability results for an SPRT calibrated at AMS four times versus another SPRT. The results are shown in terms of differences at 0°C and 300°C. The calibrations were performed one day apart. The maximum difference seen is 0.008°C confirming the statement that the repeatability of our calibration process is better than 0.01°C.

Repeatability results for six stable and six unstable RTDs calibrated on two consecutive days are given in Table 16.6. These results represent the maximum difference in the range of 0 to 300°C. Prior to checking for repeatability, these RTDs were tested for stability by a series of calibrations performed over a period of a few months. It is apparent that the repeatability error is not only dependent on the repeatability of the calibration process, but also on the repeatability of the RTD itself. The average repeatability error ranges from about 0.01 to 0.1°C for the RTDs tested here.

16.4 Self Heating Error

Measurement of RTD resistance requires a measuring current of about 1 milliampere. The current produces Joule heating (I^2R) in the sensing element and results in a temperature error. The error depends on the self heating index of the RTD. The self heating index (SHI) is a measure of changes in resistance per unit electric power generated in the RTD element from the application of electric current. The self heating index is expressed in terms of ohms per Watt

TABLE 16.4

Hysteresis Repeatability of Old RTDs

Tag	Hysteresis Error (°C)			Drift (°C)
	7/88	1/89	2/89	
3	0.25	0.25	0.25	0.1
4A	0.34	0.34	0.34	0.1
4C	0.32	0.33	0.31	2.0
5A	0.35	0.31	0.33	2.0
5C	0.35	0.36	0.36	1.0
7	0.08	0.08	0.08	0.1
8A	0.27	0.27	0.26	2.0

Above hysteresis errors all occurred at a temperature of about 145°C.

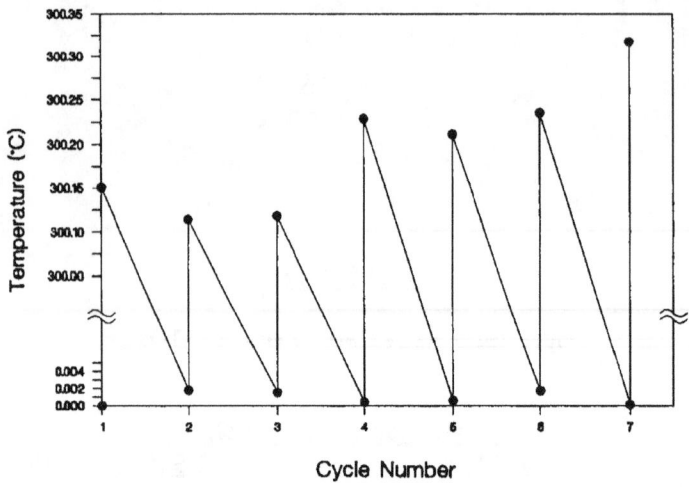

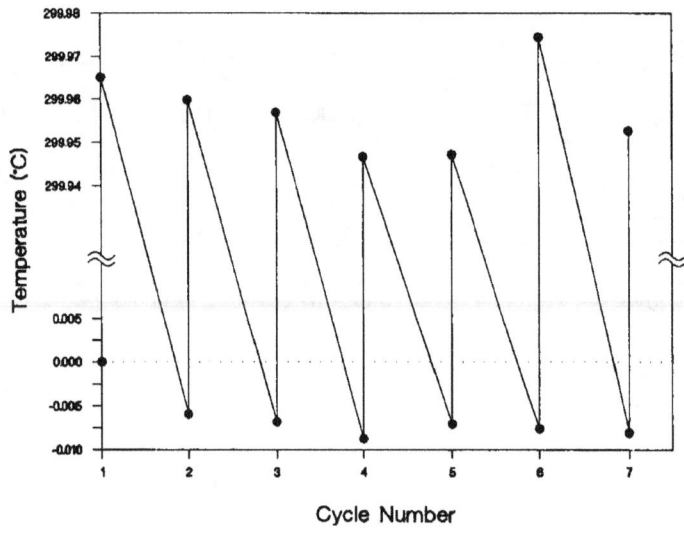

Figure 16-2.   Hysteresis Test Results at Ice Point and 300°C for Two RTDs.

TABLE 16.5

Repeatability of SPRT Calibration

| Calibration | Repeatability (°C) | |
	Ice Point	300°C
2 vs. 1	0.002	0.006
3 vs. 1	0.000	0.006
4 vs. 1	0.003	0.008
3 vs. 2	0.002	0.001
4 vs. 2	0.000	0.002
4 vs. 3	0.003	0.001

Above results represent the differences between four calibrations performed one day apart.

TABLE 16.6

One Day Calibration Repeatability

Tag	Difference (°C)
Stable RTDs	
15A	0.009
15C	0.011
16A	0.004
16C	0.005
17A	0.008
17C	0.009
Unstable RTDs	
68	0.039
61	0.053
54	0.024
62	0.138
63	0.160
57	0.021

Above differences are the maximum found in the range of 0 to 300°C.

(Ω/W). Self heating error results when RTDs are calibrated in one environment and used in another environment where the RTD may have a different self heating index. For example, RTDs are often calibrated in an oil bath and subsequently used in flowing water. An RTD may have significantly different SHIs in these two conditions.   Table 16.7 shows self heating errors measured for five 200 ohm RTDs calibrated in an oil bath inside a copper block at 300°C and used in room temperature water at 1 meter/second. The errors which are due to SHI differences range from 0.002 to 0.004°C for 1 milliampere of current. For 2 milliamperes, the errors will be four times as much.   Self heating errors are particularly pronounced when calibration is performed in an oven or a fluidized sand bath because heat transfer in these environments is very poor. The self heating error may also be large in an equalizing block such as a copper block used in the oil bath during calibration. This is the one disadvantage of equalizing blocks. Another is the long delay required for the copper block to change temperature.

Experience has shown that the average self heating indices of various RTDs are different by a factor of about 2 depending on the RTD and the conditions in which it is used. Therefore, the range for net self heating errors is 0.001 to 0.008°C.

16.5 Interpolation and Fitting Errors

The discrete resistance versus temperature points obtained during calibration must be fit to a polynomial to provide resistance versus temperature data for the temperatures in between calibration points.   Any difference between a measured point and the corresponding point calculated from the fit is called interpolation error or fitting error. The interpolation errors depend on the number of calibration points, the uncertainties at each calibration point, and the order of the polynomial. At least three calibration points and a second order polynomial are needed for generating a resistance versus temperature table for an RTD.   The most commonly used polynomial for RTD calibration is the Callendar equation which is a second order polynomial.

If more than three calibration points are available, the data can still be fit to a second order polynomial. Generally, more points and reasonably high order polynomials should result in smaller errors. Table 16.8 shows the differences between twelve point and four point calibrations of six RTDs. The results are presented in terms of differences at 0°C, 150°C, 250°C, and 300°C. The data for both cases were fit to the Callendar equation. The differences indicate that there is about 0.01°C error in using four points instead of 12 points in this case.

The twelve point calibration data were also used to determine the benefits of higher order polynomials. The results are shown in Table 16.9 and in Figure 16.3 for five RTDs. It is apparent that higher order polynomials can help reduce the error by as much as 0.006°C. The user must decide whether this gain is justified for the additional effort required to obtain more calibration points needed for the higher order fits. The Callendar fit for the data in Figure 16.3 is for four points and not for 12 points.

TABLE 16.7

Errors Due to Differences
in Self Heating Indices

Tag	Self Heating Index (Ω/W) Oil	Water	Net Error (°C)
9A	14.5	4.4	0.004
9C	14.4	4.4	0.004
16A	26.8	25.3	0.003
16C	26.2	24.8	0.002
19	7.8	2.0	0.002

Above results are from self heating tests in an oil bath at 300°C and room temperature water flowing at 1 meter per second.

TABLE 16.8

Differences Between Four Point and
Twelve Point Calibration

	Difference (°C)			
Tag	0°C	150°C	250°C	300°C
15A	0.004	0.005	0.007	0.006
15C	0.004	0.005	0.006	0.005
16A	0.004	0.004	0.007	0.007
16C	0.003	0.003	0.005	0.006
17A	0.003	0.005	0.005	0.004
17C	0.004	0.005	0.007	0.006

TABLE 16.9

Calibration Differences as a Function
of the Order of Fitting Equation

Tag	Callendar	3rd Order	Difference (°C) 4th Order	5th Order	6th Order	7th Order
15A	0.0060	0.0038	0.0014	0.0009	0.0002	0.0014
15C	0.0059	0.0040	0.0011	0.0007	0.0002	0.0002
16A	0.0052	0.0015	0.0009	0.0005	0.0001	0.0003
17A	0.0049	0.0041	0.0007	0.0004	0.0003	0.0015
17C	0.0050	0.0031	0.0009	0.0002	0.0004	0.0010

Above are the differences between the calculated and actual temperature at 300°C. For Callendar equation, four points (0, 100, 200, and 300°C) were used. For the higher order fits (3rd to 7th) twelve points were used in the fitting. These points were 0, 100, 120, 140, 160, 180, 200, 220, 240, 260, 280, and 300°C.

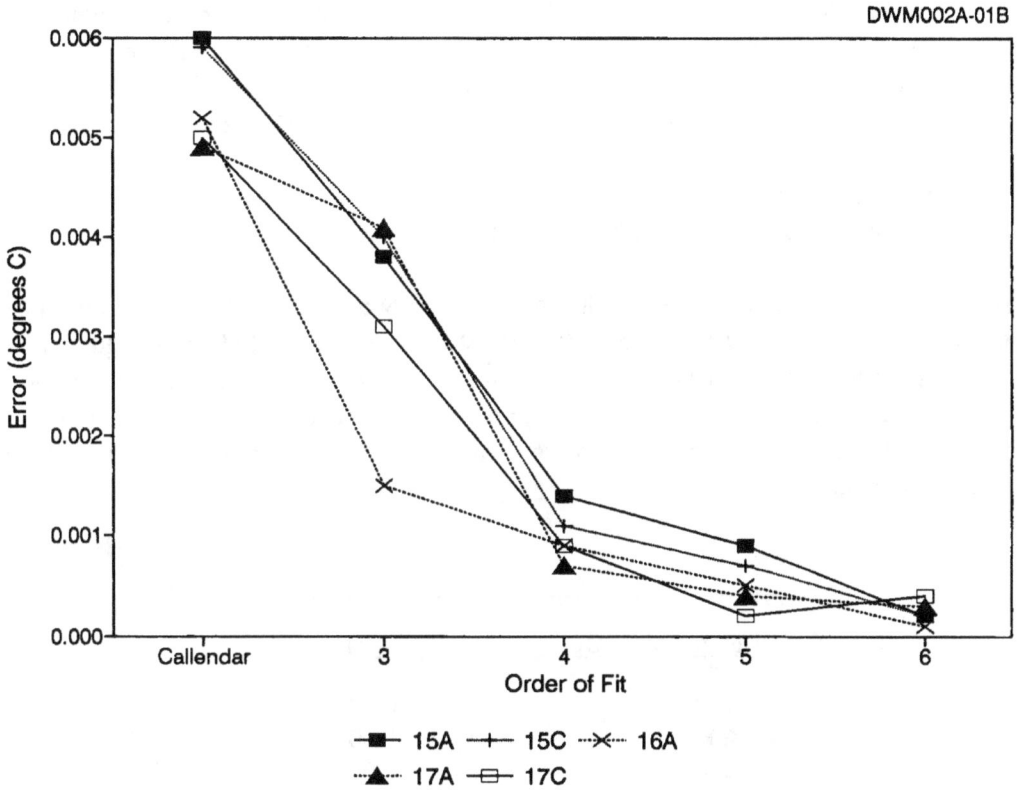

Figure 16-3.   Fitting Errors as a Function of Order of
Polynomials for Five RTDs.

For further study of fitting errors, we analytically determined the best order for fitting of data for seven RTDs. The results showed that sixth to eighth order polynomials are best depending on the RTD. We then used the best order fit for each RTD to determine the range of fitting errors encountered at 0, 160, 260, and 300°C using either the Callendar equation or the best fit possible. The results are shown in Table 16.10 in terms of the difference between the measured temperatures and the temperatures calculated from the fit. Overall, we determined the average fitting errors to have a range of 0.005 to 0.01°C based on analysis of all the data produced here. The lower end of this range is achieved with an optimum order for the fitting equation and appropriate number of calibration points, and the higher end is achieved with the Callendar equation and four calibration points (0, 100, 200, 300°C). In another attempt to quantify fitting errors, a five point calibration was performed at 0, 100, 200, 300, and 350°C on ten RTD elements. The data were fit to the Callendar equation and the results at 350°C from the equation were compared with measurements made at 350°C to determine the fitting errors at this temperature. The results are given in Table 16.11.

The uncertainty of the R vs. T curve in between any two points depends on the individual uncertainties in each of the two points. We have shown, using the Callendar equation as the model and a propagation of error analysis, that the interpolation errors can be determined given the uncertainties in the calibration points. Figure 16.4 illustrates the root sum squared (RSS) error of an RTD throughout a 0 to 300°C range. This graph assumes the following uncertainties in a 4 point calibration:

Calibration Point(°C)	Uncertainty (°C)
0	0.003
100	0.006
200	0.007
300	0.010

Note in Figure 16.4 that the uncertainties in between each pair of calibration points are smaller than the individual uncertainties of the two calibration points.

16.6 Combining the Errors

Table 16.12 summarizes the range of errors associated with a typical RTD calibration. The lower of the two values given in Table 16.12 for each source corresponds to the best that can be reasonably achieved with state-of-the-art equipment and procedures. The high end of the range is typical for a routine calibration as opposed to a sophisticated precision calibration. These errors may be systematic or random. Systematic errors are combined by calculating the

TABLE 16.10

Fitting Errors Measured for Seven RTDs

Tag	Measured Temperature (°C)	Calculated Temperature (°C) Best Fit	Calculated Temperature (°C) Callendar	Fitting Error (°C) Best Fit	Fitting Error (°C) Callendar
		0°C			
15A	0.00432	0.00428	-0.00116	0.00004	0.00548
15C	0.00442	0.00424	-0.00086	0.00018	0.00528
16A	0.00502	0.00484	-0.00045	0.00018	0.00547
17A	0.00442	0.00443	-0.00001	0.00001	0.00443
17C	0.00442	0.00425	-0.00030	0.00017	0.00472
95	0.00452	0.00452	0.00097	0.00000	0.00355
48	0.00442	0.00449	-0.00345	0.00007	0.00787
		160°C			
15A	160.41520	160.41816	160.40674	0.00296	0.00846
15C	160.41489	160.41836	160.40701	0.00347	0.00788
16A	160.41293	160.41159	160.40605	0.00134	0.00688
17A	160.41252	160.41197	160.40229	0.00055	0.01023
17C	160.41603	160.41243	160.40401	0.00360	0.01202
95	160.41365	160.41316	160.40987	0.00049	0.00378
48	160.41592	160.41922	160.40684	0.00330	0.00908
		260°C			
15A	260.38031	260.37698	260.36971	0.00333	0.01060
15C	260.38063	260.37766	260.37142	0.00297	0.00921
16A	260.38286	260.38244	260.37262	0.00042	0.01024
17A	260.38861	260.38648	260.38068	0.00213	0.00793
17C	260.38041	260.37737	260.36963	0.00304	0.01078
95	260.38467	260.38476	260.37469	0.00009	0.00998
48	260.37477	260.37544	260.35894	0.00067	0.01583
		300°C			
15A	300.39803	300.39737	300.40406	0.00066	0.00603
15C	300.39081	300.39038	300.39666	0.00043	0.00585
16A	300.39965	300.39962	300.40482	0.00003	0.00517
17A	300.39641	300.39612	300.40133	0.00029	0.00492
17C	300.39857	300.39793	300.40358	0.00064	0.00501
95	300.39178	300.39181	300.39572	0.00003	0.00394
48	300.39404	300.39424	300.40273	0.00020	0.00869

TABLE 16.11

Fitting Errors at 350°C

Tag	Temperature (°C)		Difference (°C)
	Measured	5 Point Fit	
03	350.659	350.660	0.001
9A	350.651	350.651	0.000
9C	350.640	350.639	0.001
13A	350.645	350.641	0.003
13C	350.639	350.630	0.009
15A	350.636	350.644	0.008
15C	350.632	350.641	0.009
16A	350.630	350.640	0.010
16C	350.621	350.630	0.009
21	350.618	350.577	0.040

KMP011A-04A

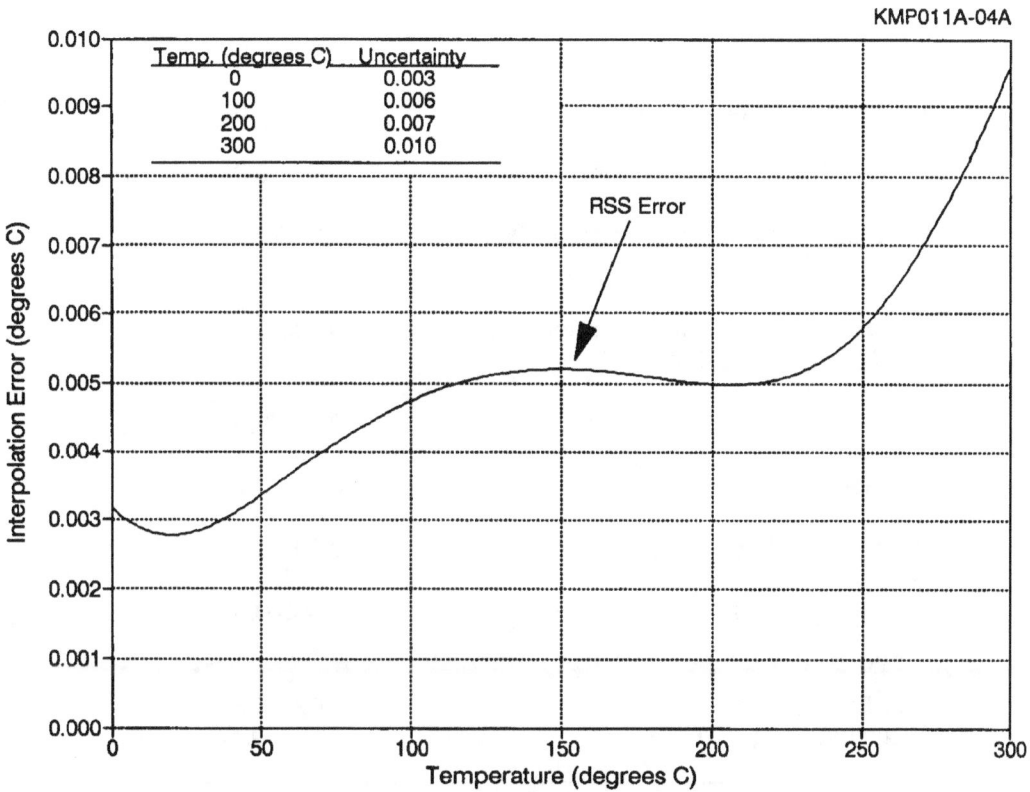

Figure 16-4. Interpolation Error.

- 103 -

TABLE 16.12

Combining the Errors

Source	Range (°C)
1. Bath Stability and Uniformity	0.009-0.02
2. SPRT Accuracy and Drift/Yr.	0.005-0.01
3. Measurement Equipment for SPRT Accuracy and Drift	0.005-0.01
4. Measurement Equipment for RTD Accuracy and Drift	0.005-0.01
5. Hysteresis	0.010-0.30
6. Repeatability	0.010-0.10
7. Self Heating	0.001-0.01
8. Fitting Errors	0.005-0.01

RSS Error (°C)	0.02 - 0.32
Maximum Error (°C)	0.05 - 0.47

algebraic sum of the individual errors and random errors are combined by calculating the square root of the sum of the individual errors squared. This is usually referred to as root sum squared or RSS error.

To obtain a range for the total errors, we first assume that all errors are random. Using the lower end numbers in Table 16.12, the RSS error is 0.02°C. This is the lowest possible uncertainty, i.e., the best accuracy that can be achieved in calibration of an RTD. To obtain the higher end of the total error, we summed all the higher end errors (assuming that they are all systematic). The result is 0.47°C. That is, the accuracy of a newly calibrated RTD will be in the range of 0.02 to 0.47°C depending on the RTD and how well it is calibrated. This accuracy holds for the entire range of 0 to 300°C because the interpolation errors cannot result in larger uncertainties than those of the calibration points. Overall, based on examination of all the results in this project and the available data in literature, it appears that the achievable accuracy in the initial calibration of an industrial RTD has a range of 0.05 to 0.1°C.

Note that in our treatment of errors, we neglected stem loss, self heating, and hysteresis of the SPRT. These errors are small if the SPRT is properly maintained and used.

17. EXTRAPOLATION ERRORS

Extrapolation errors are of concern when calibration data are extended beyond the highest temperature in which the RTD is calibrated. Nuclear plant RTDs are sometimes calibrated to 300°C and the calibration table is extended to 350 or 400°C by way of extrapolation. Table 17.1 shows the errors identified in extrapolation from 300 to 350°C for nine nuclear grade RTD elements. This was done by extrapolating a calibration which was performed up to 300°C to 350°C and comparing the results with actual measurements made at 350°C. The RTDs were calibrated at 0, 100, 200, 300, and 350°C. The first four points were fit to the Callendar equation and the resulting calibration was extrapolated to 350°C. The average error is about 0.04°C. Similar data is shown in Table 17.2 for a calibration to 200°C which was extrapolated to 350°C. This calibration was performed at 0, 100, and 200°C. The average extrapolation error is 0.2°C, five times larger than that shown in Table 17.1.

Generally, extrapolation errors can be estimated by propagation of uncertainties at the calibration points. A graph of extrapolation errors is shown in Figure 17.1 for the following ideal case in which the uncertainties of calibration points are very small.

Calibration Point (°C)	Uncertainty (°C)
0	0.001
100	0.002
200	0.003
300	0.004

The graph shows that the extrapolation errors increase significantly as extrapolation temperature is increased.

TABLE 17.1

Errors for Extrapolation from 300°C to 350°C

Tag	Temperature (°C) Measured	Extrapolated	Extrapolation Error(°C)
03	350.659	350.662	0.003
9A	350.651	350.652	0.001
9C	350.640	350.640	0.000
13A	350.646	350.629	0.016
13C	350.639	350.605	0.034
15A	350.636	350.670	0.034
15C	350.632	350.670	0.038
16A	350.630	350.669	0.039
21	350.618	350.458	0.160

TABLE 17.2

Errors for Extrapolation from 200°C to 350°C

Tag	Temperature (°C) Measured	Temperature (°C) Extrapolated	Extrapolation Error (°C)
03	350.660	350.481	0.179
9A	350.651	350.507	0.145
9C	350.640	350.447	0.194
13A	350.646	350.443	0.203
13C	350.640	350.347	0.292
15A	350.636	350.748	0.112
15C	350.632	350.758	0.127
16A	350.630	350.783	0.153
21	350.618	350.208	0.410

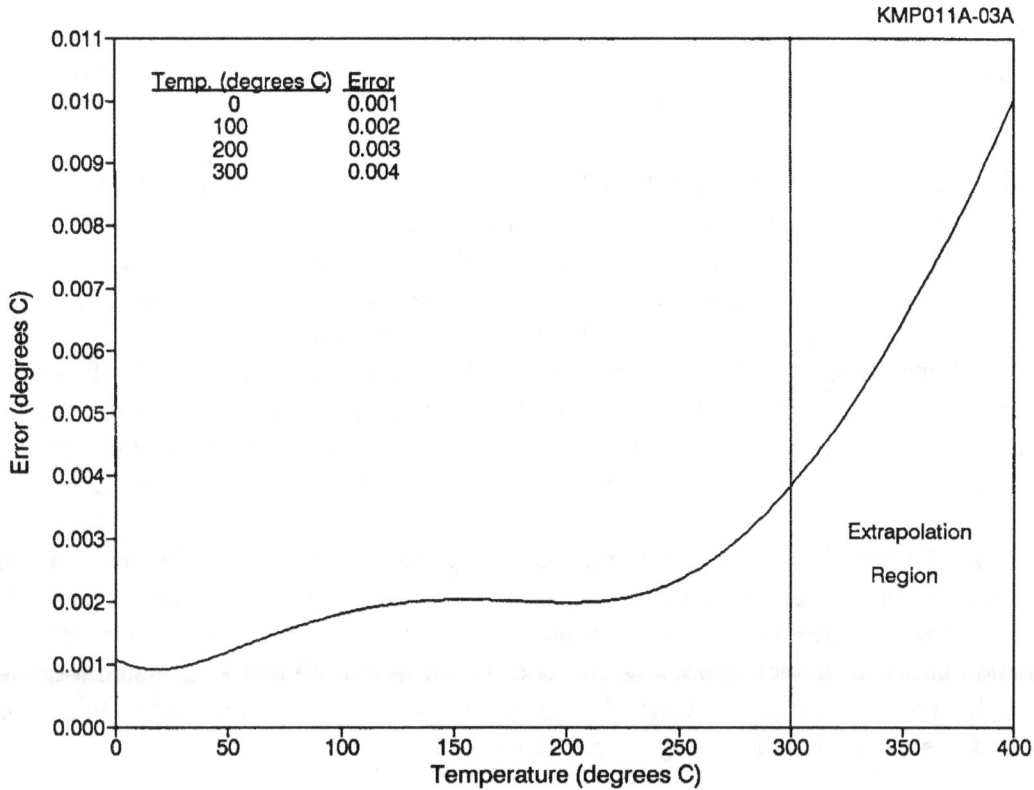

Figure 17-1. Errors for Extrapolation From 300 to 400°C
for an Ideal Calibration.

18. INSTALLATION EFFECTS ON ACCURACY

A number of installation factors can affect the accuracy of RTDs. These effects are summarized in Table 18.1 and described below.

18.1 Stem Loss

In a normally operating PWR, the primary coolant RTDs are at about 320°C, but a portion of the stem and the head assembly is outside the primary coolant pipe at a temperature of about 40 to 50°C (Figure 18.1). Generally, the stem outside the pipe is insulated, but the head assembly is not. The resulting temperature gradient causes the RTD to indicate a temperature lower than the process. This error is called stem loss and is important because it results in a nonconservative temperature indication.

Stem loss errors were demonstrated here in a quantitative and a qualitative experiment. The quantitative test involved measuring the temperature error as a function of the immersion length of an RTD in an oil bath at 300°C. The portion of the RTD which was not immersed was exposed to room temperature air (at approximately 40°C) above the oil bath. This arrangement was intended to simulate the immersion of the RTD in the primary coolant. The results are shown in Figure 18.2 and Table 18.2. The RTD was tested with and without a thermowell. Note that the immersion errors are larger for the RTD when it is in the thermowell. These results are for an RTD that is about 40 cm long with a sheath diameter of about 0.6 cm and a thermowell with an outside diameter of about 1 cm. The sensing element of this RTD is about 4 cm long.

Short RTDs that have massive thermowells or housings can only be calibrated accurately if the housing is fully immersed or well insulated to reduce the temperature gradient. Such RTDs are susceptible to large stem loss errors when they are installed in the plant. Figure 18.3 shows immersion errors for a short (about 1 cm) and a long (about 10 cm) RTD tested together at 300°C. The results are given in terms of the percentage of the length immersed. Note that the error for the short RTD is much larger than the long RTD.

The qualitative test used to demonstrate stem loss errors involved cooling the head of a 40 cm long RTD while 20 cm of it was at 300°C. The results are shown in Figure 18.4 in terms of the normalized changes in RTD output and the temperature measured inside the RTD connection head during the tests. The peak-to-peak change is about 0.2°C when the head was cooled with freon spray to around 0°C. Of course, such large temperature gradients are not involved in PWR applications and the errors are therefore much smaller. However, it is important to note that stem loss errors are often present and their magnitudes depend on the temperature gradient, the RTD diameter, heat transfer characteristics of the medium in which the RTD is used,

TABLE 18.1

Examples of Potential Installation Errors

Error	Cause
Stem Losses	Large temperature differences at two ends of RTD
EMF Errors	Dissimilar metals in large temperature gradients
Lead wire Effects	Difference in lead wire resistances in 3 wire RTDs
Insulation Resistance	Low element to sheath resistance (shunting error)
Contact Resistance	Bad or loose connectors in RTD circuit
Stratification	Temperature gradient in fluid or gas streams

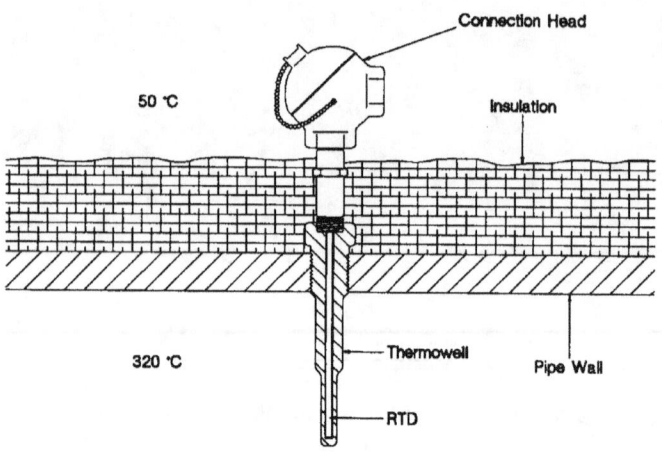

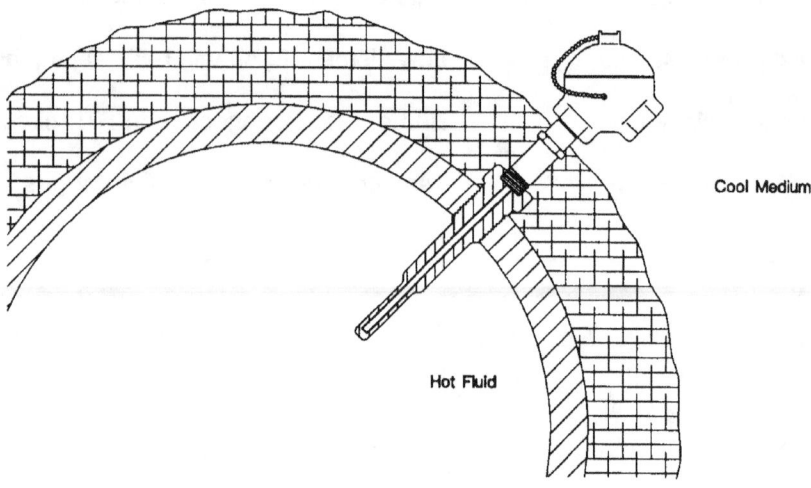

Figure 18-1. RTD Installation in Relation to Stem Loss.

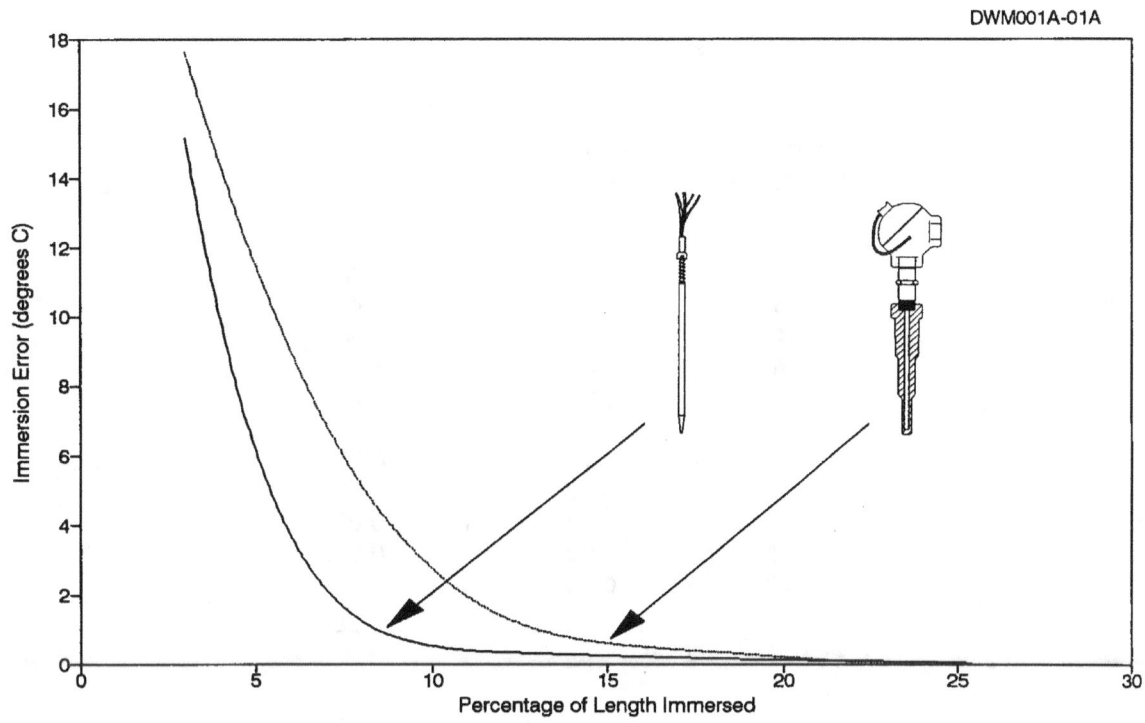

Figure 18-2.   Immersion Errors for a Well-Type RTD
With and Without Thermowell.

TABLE 18.2

RTD Immersion Error

Immersion Length (cm)	Immersion Error (°C)	
	Bare	Well
1	15.19	17.64
2	6.67	11.95
3	1.91	6.48
4	0.64	3.34
5	0.37	1.77
6	0.28	0.76
7	0.21	0.47
8	0.13	0.25
9	0.10	0.07
10	0.05	0.05
11	0.00	0.01
12	0.00	0.00

Bare: RTD without thermowell.

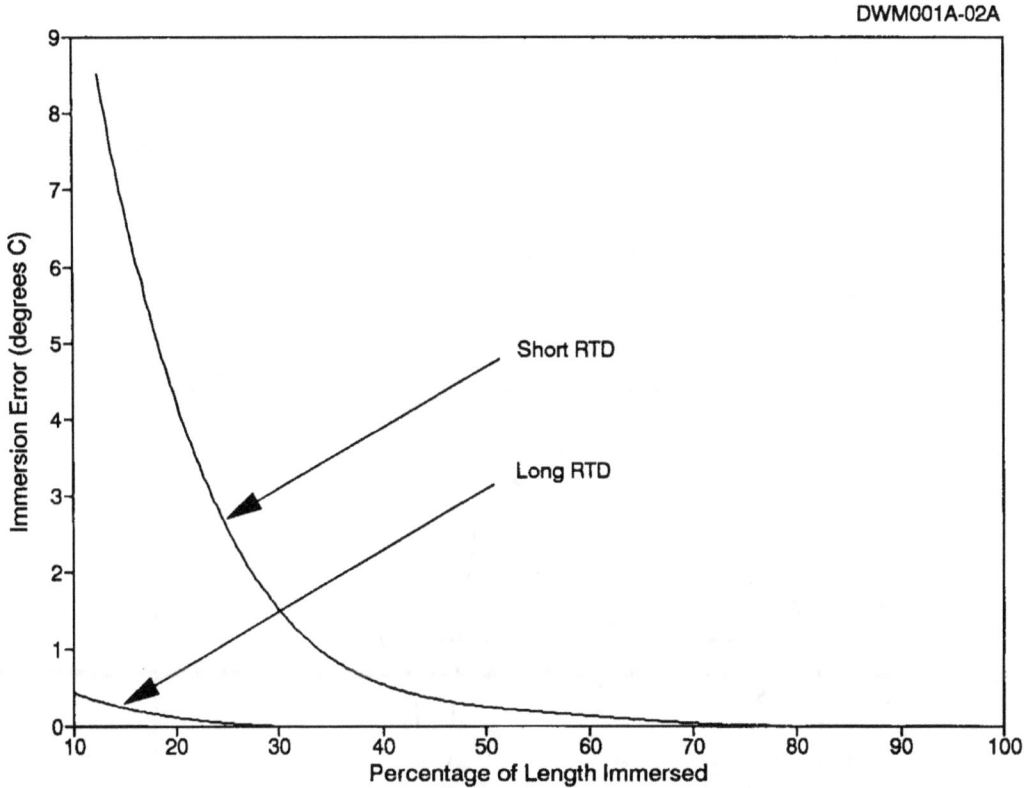

Figure 18-3. Immersion Errors for a Short and a Long RTD.

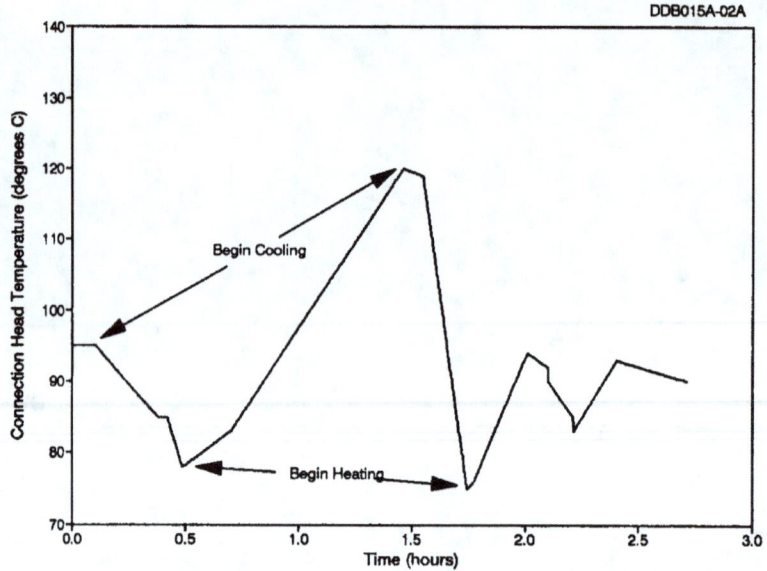

a) Temperature Inside the RTD Connection Head.

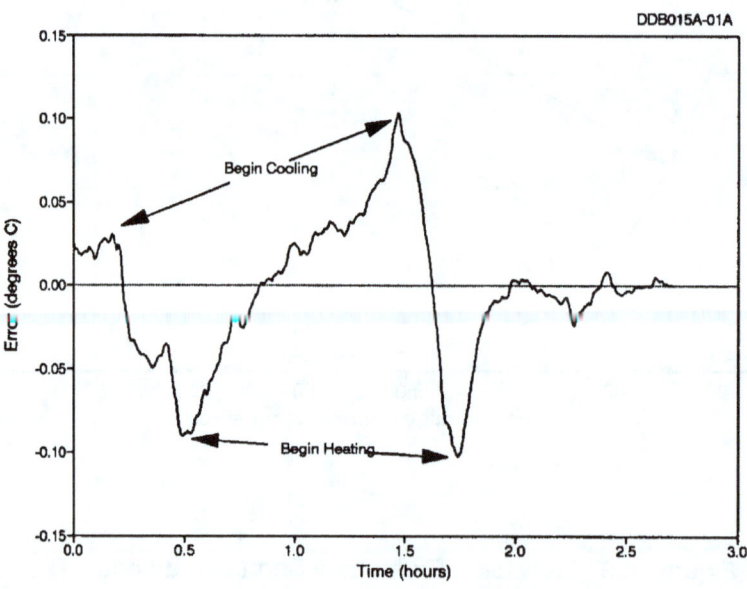

b) Temperature Indicated by RTD.

Figure 18-4. Demonstration of RTD Stem Losses.

the thermal conductivity of the RTD material, and the size of the connection head compared to the rest of the RTD. These factors can be quantified for a given installation and the stem loss error calculated. A remedy for minimizing the stem loss error is to insulate the primary coolant pipe around the RTD as was shown in Figure 18.1 to avoid a sharp temperature gradient.

18.2 EMF Errors

Thermoelectric effects result if dissimilar metals in the RTD circuit are at different temperatures. The resulting voltage (EMF) interferes with resistance measurement and can cause as much as 0.5°C error depending on the RTD, the application, and the temperature in which it is used. The error changes with time and temperature and is more pronounced at high temperatures and with shallow immersion depths. The situation is much like an RTD in series with a thermocouple. To eliminate EMF errors, the extension wires in high quality RTDs are made of platinum to minimize thermoelectric effects. During RTD calibration, AC currents are sometimes used to eliminate EMF errors.

To check for EMF, the RTD resistance is measured, the leads are reversed, and the measurement repeated. If the results are different, thermoelectric effects are present in the RTD. The correct resistance will be the average of the two measurements.

Table 18.3 shows the results of measurement of EMF effects at 300°C for some of the nuclear grade RTDs used in this project. Note that both the EMF measurements and the resistance differences (ΔR) are very small. The largest resistance difference is 0.055 ohm for a 100 ohm RTD which corresponds to approximately 0.14°C error. These measurements were also made at 0°C. On average, the EMF values at 0°C were less than 10 microvolts and ΔR values were less than 0.1 milliohm. Note in Table 18.3 that contrary to what one may expect, there is not a correlation between the EMF and the corresponding ΔR values even though ΔR is caused by EMF. The EMF versus ΔR relationships are different in different RTDs.

18.3 Lead Wire Effects

Industrial RTDs are supplied in either three or four-wire configurations to allow compensation for lead wire resistances. The compensation is straight forward for four-wire RTDs. However, for three-wire RTDs, an assumption is made that two of the three wires have equal resistances (wires 1 and 3 or 2 and 3 in Figure 18.5). Care must be taken during the installation of three-wire RTDs to ensure that the two wires are properly balanced and are not affected differently by temperature gradients. Table 18.4 gives errors resulting from lead wire imbalances in representative nuclear plant RTDs at approximately 300°C.

TABLE 18.3

Results of Measurements of EMF Effects
at 300°C in Nuclear Grade RTDs

		Resistance (Ω)		ΔR
Tag	EMF (μV)	Forward	Reverse	(Ω)
13A	0.1	429.009	429.092	-0.083
13C	2.2	428.984	428.965	0.019
3	2.0	424.765	424.787	-0.022
21	2.8	213.464	213.431	0.033
20	10.8	430.105	430.120	-0.015
11A	2.1	428.684	428.632	0.052
11C	1.0	428.636	428.657	-0.021
17A	10.4	424.857	424.844	0.013
17C	8.6	424.619	424.615	0.004
7	9.0	431.290	431.296	-0.006
16A	1.8	425.310	425.315	-0.005
16C	7.6	425.080	425.057	0.023
5A	15.5	213.493	213.438	0.055
5C	3.4	213.808	213.803	0.005
19	8.3	430.104	430.116	-0.012
9A	1.2	429.887	429.881	0.006
9C	1.3	430.018	430.012	0.006
18	3.9	432.059	432.032	0.027
15A	4.6	430.488	430.478	0.010
15C	2.8	430.689	430.680	0.009
22A	5.0	426.385	426.377	0.008
22C	6.1	426.190	426.194	-0.004
23	17.9	431.018	431.032	-0.014
24	16.7	425.260	425.305	-0.045
14	12.4	428.503	428.533	-0.030

μV = *Microvolt*   ΔR = *Forward Resistance - Reverse Resistance*

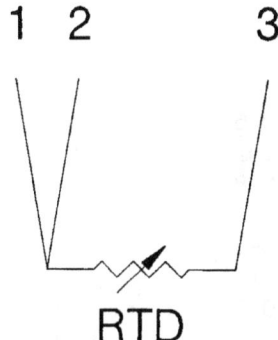

Figure 18-5. Schematic of a Three-Wire RTD.

TABLE 18.4

Lead Wire Imbalance in Three-Wire RTDs
and the Resulting Errors

Tag	Lead Wire Imbalance (Ohms)	Corresponding Error Error (°C)
1	0.248	0.348
2	0.013	0.018
3	0.047	0.065
4	0.108	0.151
5	0.009	0.012
6	0.045	0.063
7	0.102	0.143
8	0.008	0.012
9	0.095	0.133
10	0.010	0.014
11	0.069	0.096
12	0.016	0.022
13	0.070	0.098
14	0.030	0.042
15	0.061	0.085
16	0.238	0.333

18.4 Insulation Resistance

A high and stable insulation resistance is important for the accuracy of an RTD. Typically, a room temperature insulation resistance of at least 100 megohm with 100 VDC applied between any RTD lead and the sheath is desired. As the temperature increases, the insulation resistance decreases. It is important for the insulation resistance at high temperatures to be much larger than that of the RTD element. This is one reason that the ice point resistance (R_o) of some SPRTs is usually low (e.g., 25 Ω), to minimize the effects of low insulation resistance at high temperatures. Another reason is to minimize self heating errors.

A cause of RTD degradation is failure of the insulation resistance due to moisture intrusion into the sheath. A low insulation resistance causes the effective RTD resistance to be lower than normal and result in a low temperature indication. For example, for a 200 ohm RTD operating at 300°C, the indicated temperature will have a -0.002°C error if the insulation resistance is 100 megohm and a -0.2°C error if the insulation resistance is reduced to 1 megohm. Tables 18.5 and 18.6 show insulation resistances for nuclear and commercial grade RTDs measured at room temperature and at 300°C using a megohmeter at 100 VDC. Note that there is a wide variation in the changes that occur in the insulation resistances of both nuclear and commercial grade RTDs with temperature. These variations are probably due to differences in properties of the insulation materials used in different RTDs and the moisture content of the insulation materials. From the data shown in Figure 9.3, if the insulation is made of dry MgO, the insulation resistance ratio from room temperature to 300°C should be about 600.

Another effect of low insulation resistance due to moisture in the RTD is a noisy temperature signal from the RTD.

18.5 Other Installation Errors

Installation errors in addition to those discussed above include:

- Contact Resistance. Inadequate or loose connections in an RTD circuit can produce additional resistances and cause incorrect indications or a noisy temperature signal.

- Stratification Error. If a fluid is not well mixed, a temperature gradient will be encountered resulting in a temperature error. In some plants, several RTDs are used in the same plane to obtain an accurate average temperature for the process fluid. In other plants, several samples of the reactor coolant are routed to a manifold to provide a well-mixed sample for an average temperature measurement.

TABLE 18.5

Effect of Temperature on Insulation
Resistance of Selected Nuclear Grade RTDs

| Tag | Insulation Resistance (MΩ) | | Room Temp./300°C Ratio |
	Room Temp.	300°C	
3	2,000	100	20
4A	30,000	1,000	30
4C	30,000	1,000	30
5A	3,000	10	300
5C	3,000	10	300
7	20,000	30	660
8	50,000	500	100
9A	500	150	3
9C	400	150	2.6
11A	30,000	2,000	15
11C	30,000	2,000	15
12A	50,000	1,000	50
12C	50,000	1,000	50
13A	50,000	1,000	50
13C	50,000	1,000	50
14	50,000	1,000	50
15A	50,000	150	333
15C	50,000	150	333
16A	100,000	500	200
16C	100,000	500	200
17A	50,000	100	500
17C	50,000	100	500
18	50,000	20	2,500
19	100,000	10	10,000
20	.01	.01	1
21	10,000	20	500
22A	50,000	100	500
22C	50,000	100	500
23	50,000	100	500
24	15	1	15

MΩ : megohm

TABLE 18.6

Effect of Temperature on Insulation
Resistance of Selected Commercial Grade RTDs

Tag	Insulation Resistance (MΩ)		Room Temp./300°C Ratio
	Room Temp.	300°C	
26	100,000	10,000	10
27	1,000	300	3.3
35	5,000	0.7	7,100
36	1,000	50	20
38	3,000	8	375
42	1	.3	3.3
43	50,000	5,000	10
44	10,000	1,000	10
51	30,000	4,000	7
53	100,000	100	1,000
54	50,000	1,000	50
1095	50,000	10,000	5
1248	50,000	10,000	5
97A	50,000	150	333
97C	50,000	100	500
98A	100,000	800	125
98C	100,000	1,000	100
99A	5,000	40	125
99C	4,000	50	80

$M\Omega$: *megohm*

19. CALIBRATION OF RTD TRANSMITTERS

Temperature measurement accuracy with an RTD depends not only on the accuracy of the RTD, but also on the accuracy of the temperature transmitter used for conversion of resistance to equivalent temperature. The calibration of temperature transmitters often involves a "zero" and span or bias and gain adjustments. An adjustment for nonlinearity is usually not available. This results in a temperature error which peaks at the middle of the temperature range for which the transmitter is calibrated. The error can be minimized by linearizing the calibration curve for the range of interest. This may be done by fitting a straight line through the RTD resistance versus temperature curve in the temperature range of interest and using the intercept and slope of the straight line to calibrate the transmitter. A procedure for performing the linearization is:

1. Use the most current RTD calibration table and look up 10 to 20 resistance versus temperature values in the temperature range of interest.

2. Fit the data to a straight line (using a least squares fitting approach).

3. Use the slope and the intercept of the straight line to provide data to calibrate the transmitter.

Figure 19.1 shows the error curves for the range of 250°C to 350°C with and without linearization for a 100 ohm RTD with $\alpha = 0.00385$ and $\delta = 1.5$. The non-linearized curve represents the difference between the resistance versus temperature curve of the RTD and a straight line drawn between the end points of the curve and the linearized curve represents the difference between the resistance versus temperature curve of the RTD and the straight line fit throughout the curve. Note that with linearization, the errors are distributed throughout the temperature range and therefore, have a smaller maximum value. Beside reducing the errors, the advantage of linearization is that it makes the transmitter calibration easier and more uniform. It must be pointed out that plant procedures for calibration of RTD transmitters are usually written to accomplish the linearization by adjusting the zero and span repeatedly until the errors near the middle of the temperature range is reduced to below a predetermined value. The procedure suggested here accomplishes a similar result without a need for any iteration to meet the acceptance criteria.

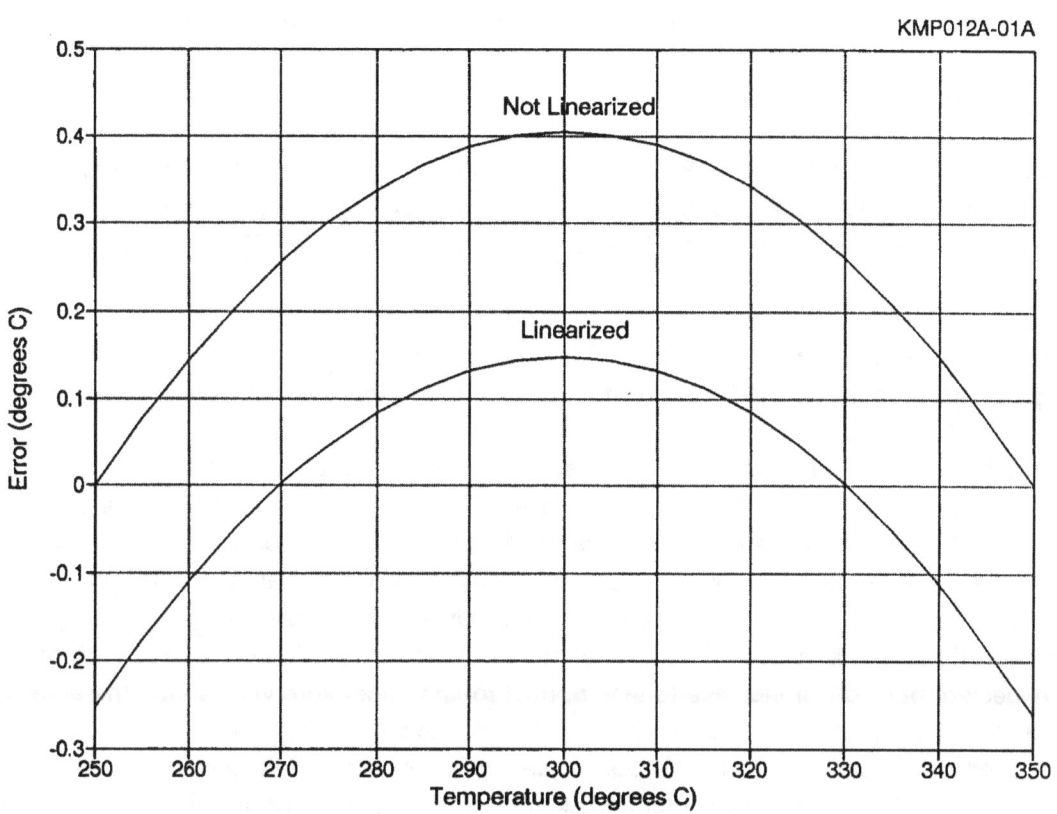

Figure 19-1.   RTD Transmitter Calibration Errors With
and Without Linearization.

20. ANNEALING

Annealing is used to correct any calibration shift due to work hardening of the sensing element. The annealing of platinum wire occurs at above 400°C. The annealing time and temperature are related. A good combination is annealing at about 450°C for a day unless the RTD is used at a higher temperature. In this case, the annealing should be performed at about 50°C above the highest temperature at which the RTD is normally used. The RTD should be cooled slowly after annealing. Removing the RTD from the annealing furnace and allowing it to cool at ambient temperature is an adequate practice. If the operating range of the RTD does not permit annealing at 450°C, then the annealing can be done at a lower temperature for a longer period of time. A side benefit of annealing is that it sometimes drives moisture out of the RTD and results in improved stability.

Annealing to reverse cold working effects in platinum wire can also be accomplished using a high electric current applied across the RTD element. The advantage of electrical annealing is that it allows in-situ annealing and it has a less significant effect on RTD materials which may not withstand high temperatures in a furnace. In addition to annealing, high electric currents can be used to heal the RTDs that have failed due to open elements. We attempted this using about 2 amperes of current to heal an RTD that had an open element. The current healed the RTD to the point that it could be used, but only for a period of four hours.

The effect of annealing was demonstrated using a few relatively unstable RTDs. The RTDs were first calibrated repeatedly for a week to establish their instability level. They were then annealed at approximately 400°C for one day and recalibrated daily for another week. The drift results before and after annealing are shown in Figure 20.1. It is apparent that the average stability of the RTDs improved. An unusual behavior was observed during annealing of one group of RTDs. The insulation resistance of the group decreased during the annealing process as expected, but did not increase to their normal room temperature values after the annealing process was completed. The data for one RTD from this group is shown in Figure 20.2 and compared with that of a normal RTD. Subsequent measurements of the insulation resistance of this group of RTDs performed daily for a week showed recovery in several of the RTDs as shown in Table 20.1. In fact, the insulation resistances of a few of these RTDs eventually exceeded the values obtained before annealing.

Annealing should be performed with caution since the construction material in some RTDs may not withstand high temperatures for a long period of time. The maximum operating range of the RTD as specified by its manufacturer must be considered before annealing.

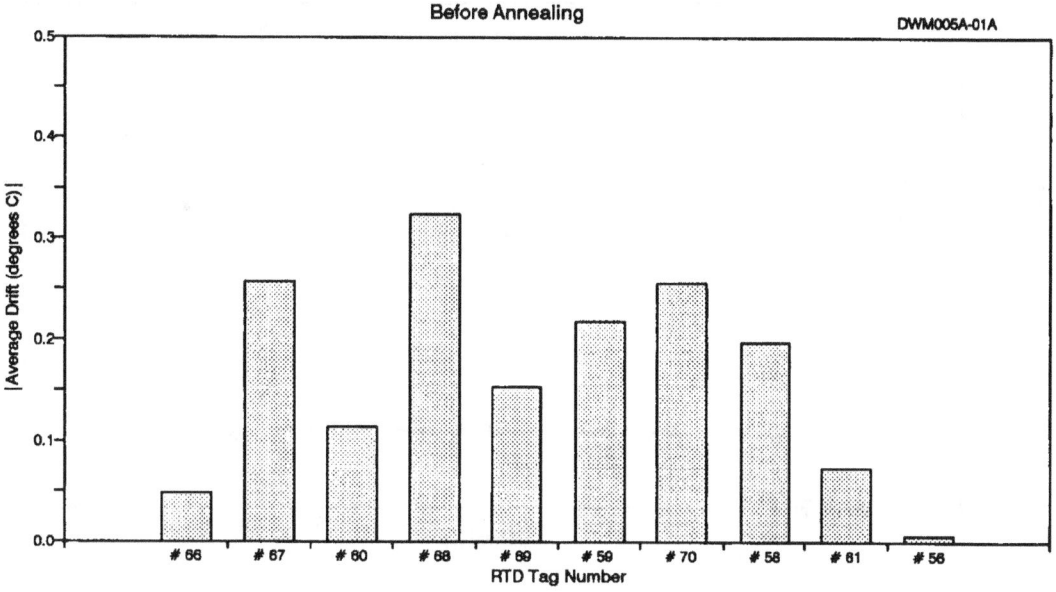

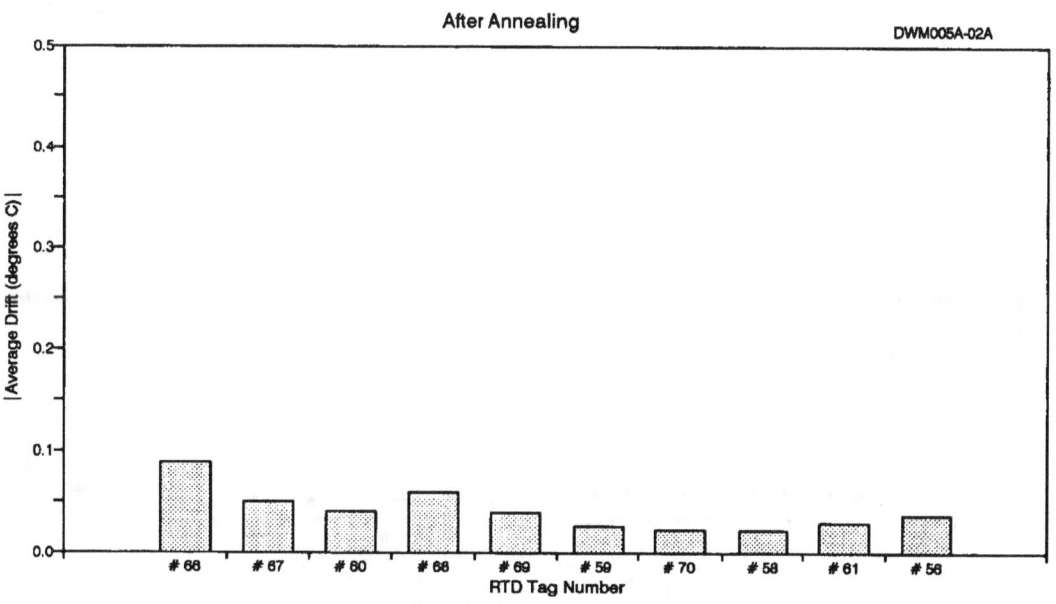

Figure 20-1. RTD Stability Before and After Annealing.

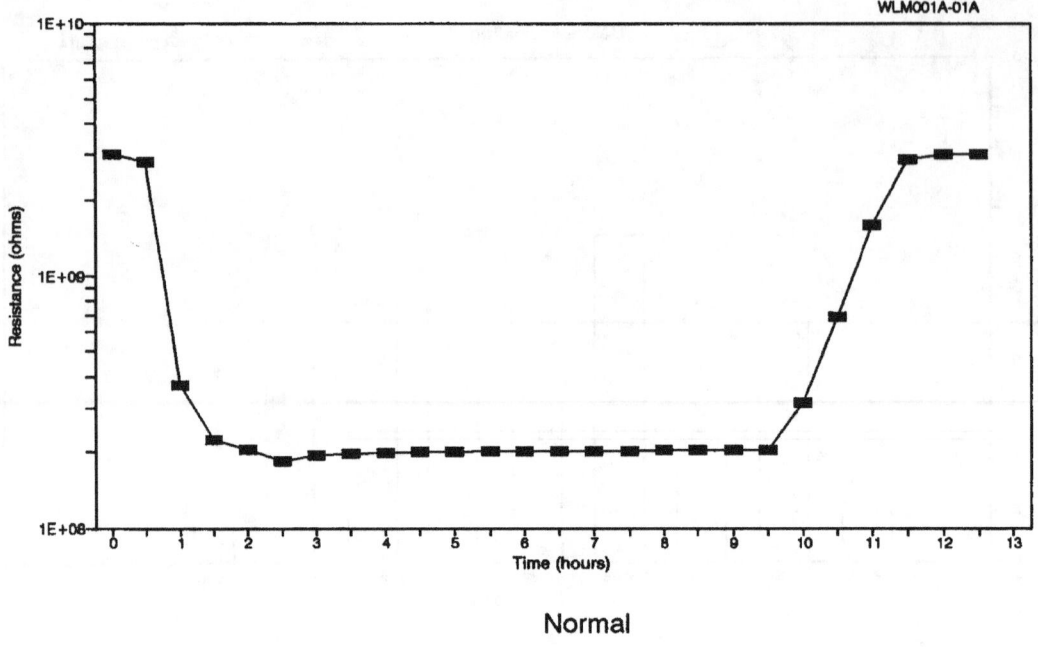

Normal

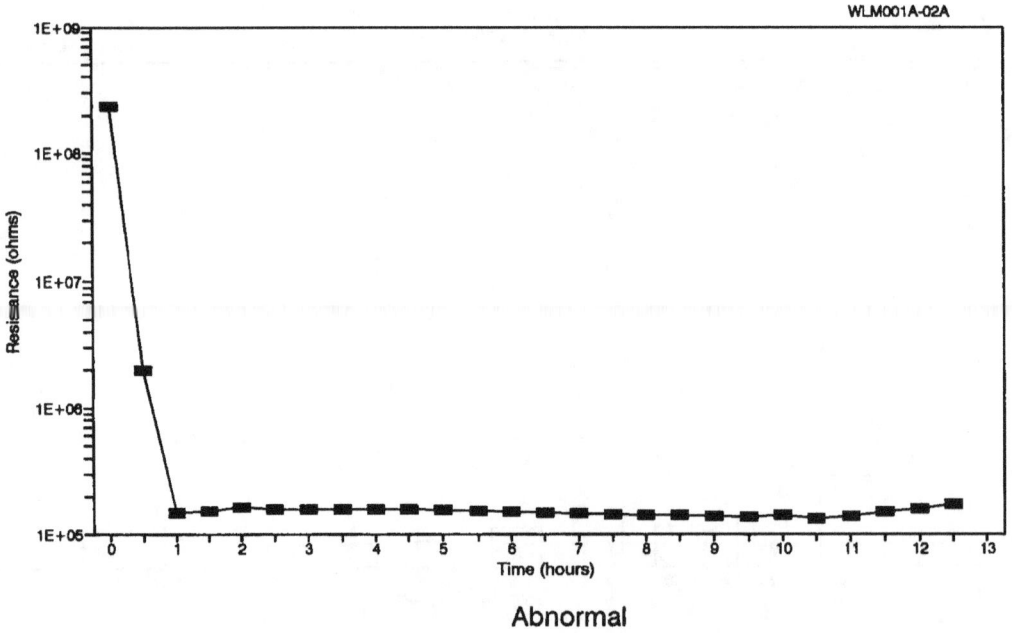

Abnormal

Figure 20-2.   Insulation Resistance Results from Measurements Made
During Annealing of a Normal and an Abnormal RTD.

TABLE 20.1

Insulation Resistance Recovery

Tag	Before Annealing	Daily Measurements After Annealing				
		Day 1	Day 2	Day 3	Day 4	Day 5
66	100	0.1	8	14	150	150
67	30	0.4	5	8	90	90
60	50	0.8	6	9	35	35
68	50	0.2	3	5	15	15
69	40	0.3	3	5	4	4
59	20	0.8	25	25	55	61
70	200	4	80	80	300	410
58	200	2	40	40	70	72
61	200	2	45	45	40	120
56	5000	8	125	125	200	250

Above measurements are in megohm at room temperature with 100 VDC applied.

Successful results of annealing of an SPRT are shown in Figure 20.3. The results are shown in terms of the resistances of the SPRT measured at the triple point of water (R_{TP}) during a three year period. Note that R_{TP} shifted by about 0.03°C in month six. This shift was successfully reversed by annealing the SPRT at 630°C for four hours.

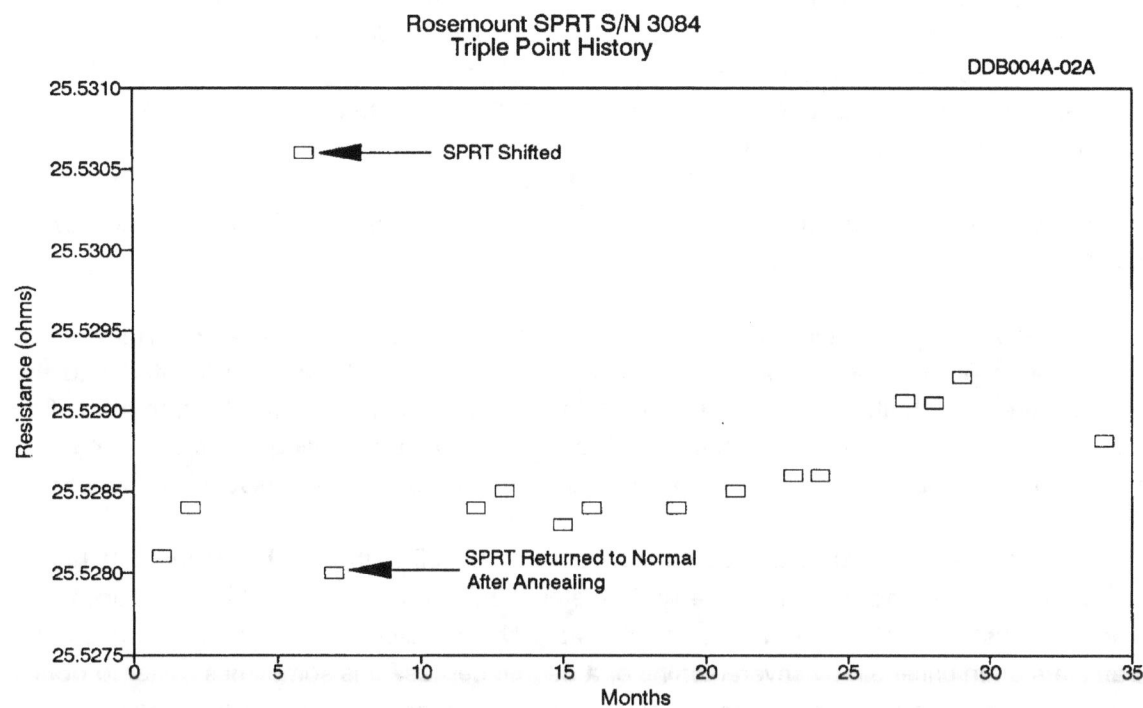

Figure 20-3. Results of Annealing of an SPRT.

21. LABORATORY CALIBRATION PROCEDURE

Proper calibration of an RTD requires a calibrated SPRT and calibrated resistance measurement equipment. The SPRT must have been calibrated by NIST and maintained properly. Before performing any calibration, the resistance of the SPRT at the triple point of water (0.01°C) must be checked to verify that the SPRT has not shifted. This should be done using a triple point cell similar to that shown in Figure 21.1. Instructions for using a triple point cell are provided in NBS Monograph 126 (Reference 4). The SPRT shall not be used if its triple point indication has shifted by more than 0.01°C. A shift in the triple point indication of an SPRT can often be corrected by annealing the SPRT at a temperature recommended by the SPRT manufacturer. After annealing, the triple point resistance of the SPRT must be checked to verify that the shift has been nulled by annealing. The SPRT can be used directly in performing the calibration or used to calibrate a secondary standard. If a secondary standard is used, the uncertainties must be increased accordingly.

For resistance measurements, a resistance bridge or a precision digital multimeter may be used. The equipment must have valid calibration traceable to NIST.

The calibration process involves making several pairs of measurements, each pair known as a calibration point. The pair includes the resistance of the RTD being calibrated and the temperature of the bath as indicated by the SPRT (Figure 21.2). A minimum of three calibration points shall be taken, two points close to each end of the RTD's operating range and one at the middle. If more than three calibration points are taken, they should be reasonably spaced.

For temperatures of 0 to 350°C, an ice bath and an oil bath may be used. For higher temperatures, a heated metal block, a liquid metal bath, or a fluidized sand bath is necessary. The user must be cautioned, however, that the stability and uniformity of these high temperature baths are often uncertain by several tenths of a degree celsius. It is sometimes better to obtain more calibration points at moderate temperatures and extrapolate to higher temperatures than to attempt to include an uncertain high temperature point. If high temperature points are reached by extrapolation, the extrapolation errors must be included in determining the accuracy of the RTD. Extrapolation should not be used to extend the RTD calibration beyond 20 percent of the highest calibration point.

The temperature of the ice bath and the oil bath must be measured with the SPRT. Unless the ice bath is properly made of distilled water and distilled ice, the ice bath temperature cannot be assumed to be 0°C. The ice is essentially used to provide a stable temperature medium rather than a medium whose temperature is known to be zero.

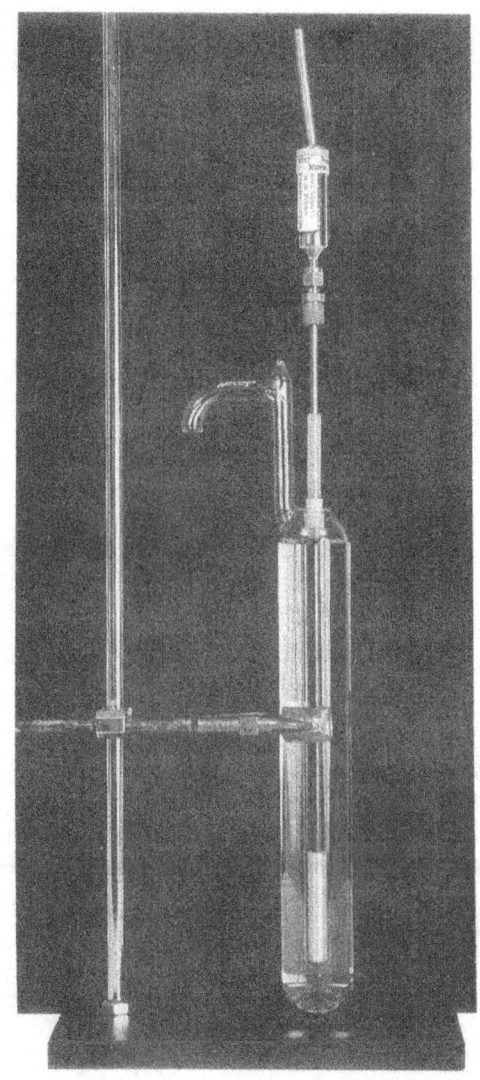

Figure 21-1.   Photograph of Triple Point Cell
                Used for Checking of SPRTs.

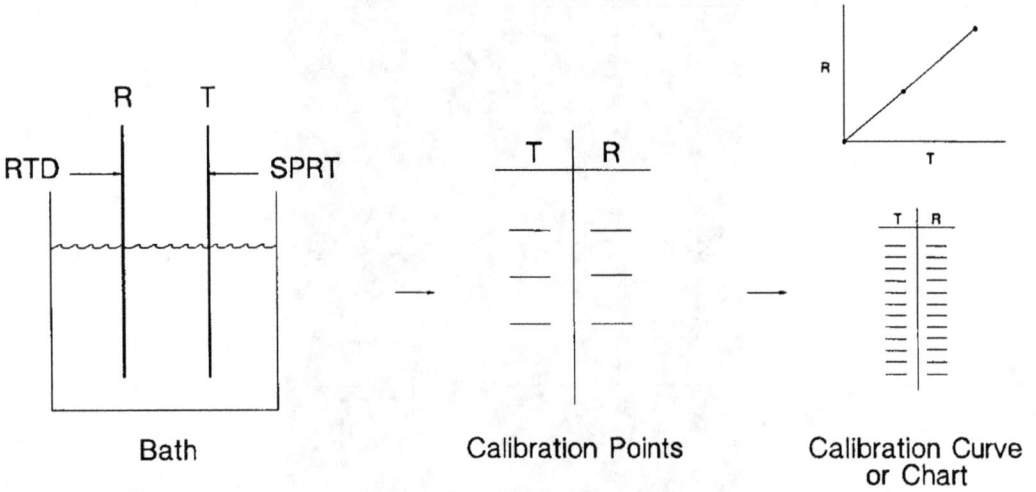

Figure 21-2. Illustration of an RTD Calibration Process.

22. ASSESSMENT OF FACTORY CALIBRATION

As we discussed in Section 16, the calibration accuracy of an RTD depends on many factors including the accuracy of the test and measurement equipment and the calibration procedure. These factors cause the results of calibrations performed in two different facilities to be at least slightly different unless the same equipment and procedures are used. Generally, a difference of less than 0.2°C indicates that the two facilities are performing equally accurate calibrations. A difference larger than 0.2°C would be an indication of differences in calibration practices between the two facilities or an indication of a shift in the RTD.

In order to assess the initial accuracy of RTDs being supplied to the nuclear industry, representative RTDs from the four U.S. manufacturers of nuclear grade RTDs were calibrated and the results were compared with the calibration provided for each RTD by its manufacturer. Figure 22.1 shows the differences between the AMS calibration and that of the manufacturers at 300°C. The results are given for fourteen RTDs representing three from each of the four manufacturers and two new RTDs contributed by a utility along with the factory calibration data. Seven of these RTDs are dual element providing a total of 21 elements that were calibrated. Except for two elements, the differences are less than about ± 0.2°C for most of the RTDs. The differences are attributed to:

- Differences between the equipment, procedures, and data processing methods.

- The repeatability of the RTDs.

- Any shift in calibration due to shipping, handling, and normal drift with time between calibrations.

A review of manufacturer's procedure for calibration of nuclear grade RTDs indicated that, in most cases, a four-point calibration is necessary. An example of the calibration points recommended by manufacturers and the associated tolerances is given in Table 22.1.

The accuracies of the commercial grade RTDs were also determined. This was done by calibrating eleven commercial grade RTDs and comparing the results with that of a DIN standard curve. DIN is the standard to which the commercial grade RTDs used here were made. The differences at 300°C are shown in Figure 22.2. DIN is a German standard for industrial RTDs. The results in Figure 22.2 show that for accuracies of better than 2°C, the RTDs must be calibrated.

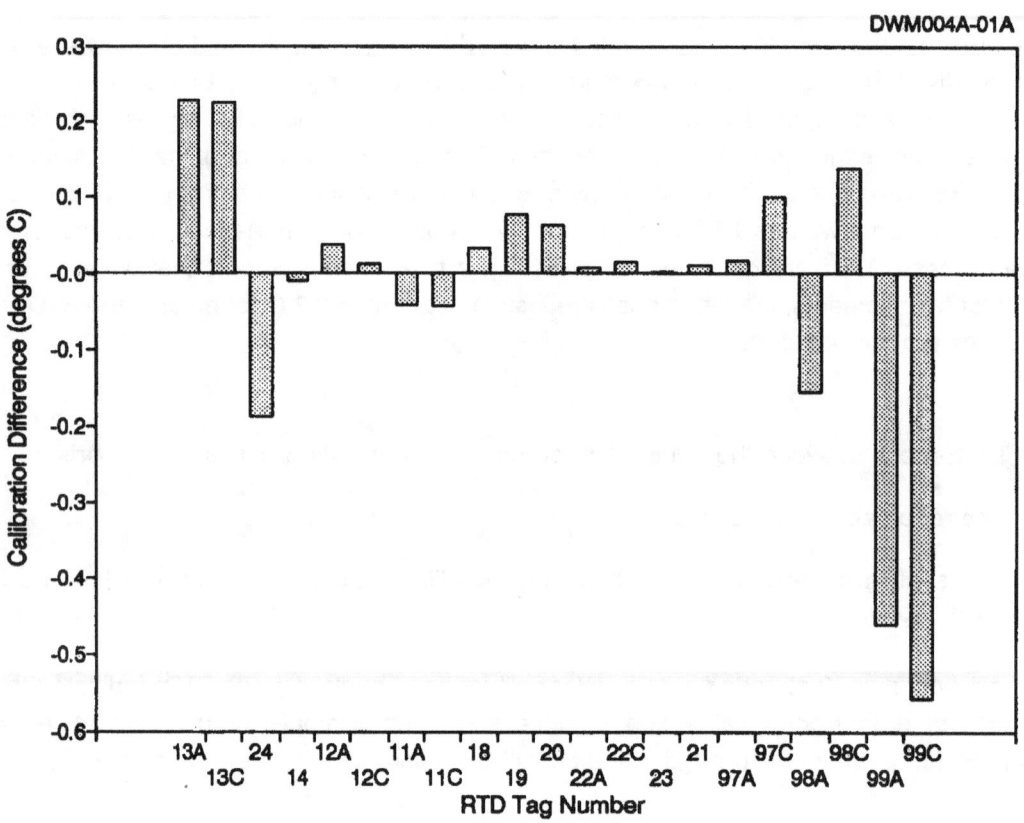

DWM004A-01A

Figure 22-1. Differences Between AMS and Manufacturer Calibration.

TABLE 22.1

Calibration Points and Tolerances Suggested
by Manufacturers for Nuclear Grade RTDs

Calibration Point (°C)	Tolerance (°C)
0	0.01-0.03
100	0.02-0.06
260	0.08
270	0.07
290	0.03
315	0.04
320	0.03
330	0.09

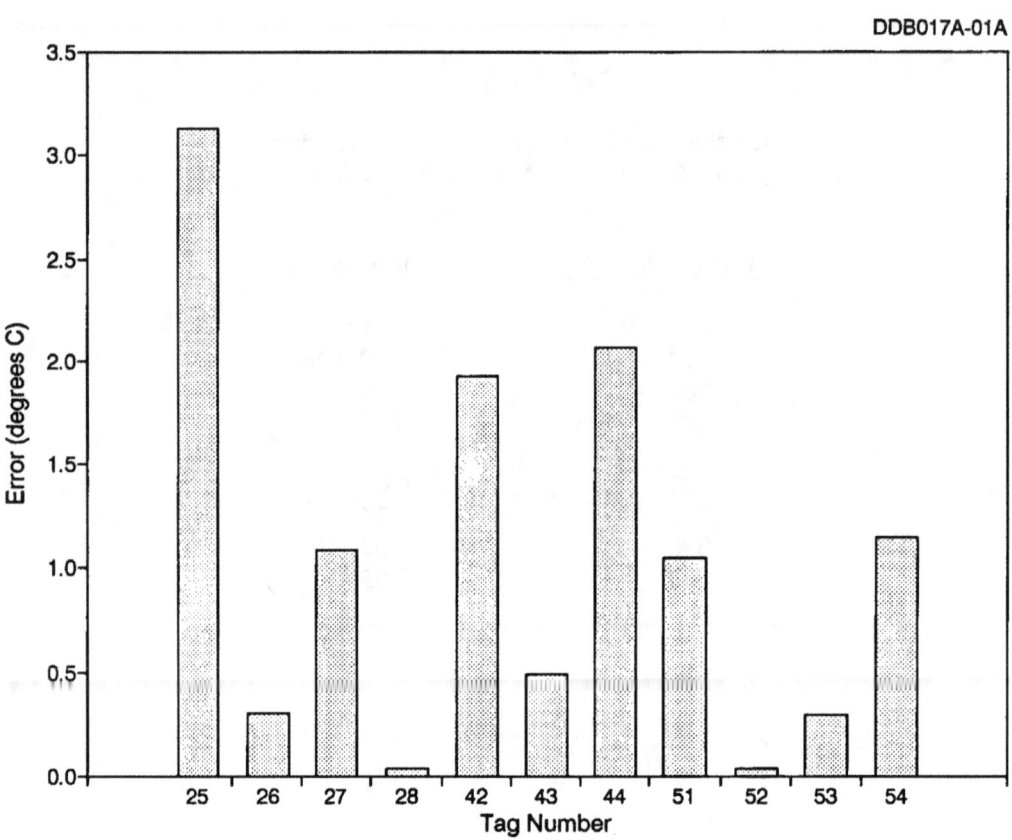

Figure 22-2. Errors in DIN Calibration of Commercial RTDs.

23. RTD CROSS CALIBRATION

23.1 Description of Test

Cross calibration is a method for on-line testing of the accuracy of installed RTDs. With the plant at isothermal conditions, the resistances of all primary coolant RTDs are measured and converted to temperature using the current calibration data for each RTD. The resulting temperatures are then averaged and the average is assumed to represent the true process temperature. This temperature is then compared with the indication of each RTD to assess the accuracy of the individual RTDs.

Table 23.1 shows the results of a cross calibration of sixteen RTDs. This was performed in a PWR using a system which scanned the RTDs four times to measure their resistances. The resulting resistances are recorded as Pass 1 through 4 and the average of the four measurements is calculated for each RTD. The average resistances are converted to temperature and the resulting temperatures are averaged. This average temperature is then compared with the temperature indicated by each RTD and their deviations are calculated. Any RTD that deviates from the average by more than a predetermined criteria is flagged or excluded from the average depending on the magnitude of the deviation. Table 23.2 provides the criteria used in seven nuclear plants for excluding (from the average) or flagging of RTDs. The RTDs that are excluded (from the average) are usually replaced and those that have been flagged are often corrected by adding a bias in the temperature transmitter to null the deviation.

Cross calibration is sometimes performed at several plateaus to verify the accuracy of the RTD over a wide temperature range. In some plants, new resistance versus temperature charts are generated using the cross calibration data. A chart generated with cross calibration data is not as reliable as a calibration chart generated with laboratory measurements. The approach is similar to performing a calibration without an ice point. The impact of calibration without an ice point on accuracy was studied in the laboratory using six nuclear grade RTDs. A four point calibration was performed at 0, 100, 200, and 300°C, and the data were analyzed with and without the ice point. The results are shown in Table 23.3 in terms of the differences at ice point and 200, 280, and 300°C. These are the differences between the fitting results with and without an ice point. The differences are small at higher temperatures and large at 0°C. This was further verified by repeating the calibration to obtain 12 points in the 0 to 300°C range. The data were analyzed using the four normal calibration points (0, 100, 200, and 300°C) and four high temperature calibration points (160, 200, 240, and 300°C). The difference in the fitted results are shown in Table 23.4 for four temperatures: 0°C, 200°C, 280°C, and 300°C. Again, the differences are small at high temperatures and large at 0°C.

TABLE 23.1

In-Plant Cross Calibration Results

Temperature Plateau: 280°C

Tag	Resistance Measurements (Ohms)				Avg. Res. (Ohms)	Temp. (°C)	Dev. (°C)
	Pass 1	Pass 2	Pass 3	Pass 4			
1	408.3075	408.3614	408.4015	408.3950	408.3663	278.070	-0.035
2	408.0508	408.1079	408.1378	408.1279	408.1061	278.243	0.137
3	408.0823	408.1500	408.1696	408.1619	408.1410	278.034	-0.072
4	408.4108	408.4825	408.4903	408.4813	408.4662	278.206	0.100
5	408.0488	408.1355	408.1286	408.1235	408.1091	278.161	0.055
6	408.2716	408.3567	408.3478	408.3457	408.3307	278.212	0.106
7	408.1202	408.1654	408.1850	408.1714	408.1605	277.759 #-0.347	
8	408.3659	408.3917	408.4105	408.4059	408.3935	278.529 # 0.423	
9	408.3339	408.1414	408.4245	408.4219	408.3986	278.047	-0.059
10	407.9606	408.0369	408.0359	408.0314	408.0162	278.053	-0.053
11	408.3245	408.3936	408.3777	408.3836	408.3698	278.061	-0.045
12	408.4128	408.4656	408.4527	408.4555	408.4466	278.124	0.018
13	408.2628	408.3532	408.3531	408.3370	408.3265	277.961	-0.145
14	407.9214	408.0120	408.0225	408.0074	407.9908	278.067	-0.039
15	408.5879	408.5910	408.5987	408.5848	408.5906	278.318	0.212
16	408.1690	408.1644	408.1806	408.1533	408.1668	277.847	-0.259

Average Temperature (°C):   278.106

 # Denotes that acceptance criteria of 0.3°C is exceeded.

TABLE 23.2

Examples of Cross Calibration Criteria
in Various Nuclear Power Plants

Nuclear Plant	No. of Repeats	No. of Plateaus	Acceptance Criteria (°C)			Remarks
			Exclude	Flag	Max. Acceptable	
1	4	1	0.6	0.17	1.1	1
2	4	1	0.6	0.17	1.1	1
3	4	4	0.6	0.11	0.6	2
4	4	2	0.3	0.30	1.1	3
5	4	4	0.6	0.27	N/A	4
6	N/A	1	0.4	0.17	N/A	5
7	4	4	0.7	0.11	0.7	6

Plant stability criteria for RTD cross calibration is typically about ± 0.3 to ± 0.6°C.

Remarks:

1. *Cross calibration data is taken for any number of plateaus. However only the data for 292°C is used to meet acceptance criteria and adjust the temperature transmitters as needed.*

2. *Data can be taken at a constant heat up rate. On sixteen RTDs, data is taken as follows: RTD number 1 to 16, reverse current 16 to 1, reverse current 1 to 16, etc. This presumably corrects for both the ramping temperature and for EMF effects (reversing the current). Data is taken around 95, 170, 230, and 292°C. The data is used to correct 2nd order fit to the data.*

3. *Two plateaus: 170°C and 292°C. For deviations greater than 0.17, the deviation at 170°C and 292°C is used to determine the error offset and the slope and apply the corrections to the transmitter.*

4. *Data is taken at four plateaus on sixteen RTDs sequentially, 1-16, 16-1, etc. The heat up rate is also measured.*

5. *The sixteen RTDs in this plant are tested one channel (four RTDs) at a time. Data is taken for 25 minutes at 5 minute intervals. This is repeated for all four channels. The plant stability requirement for the tests is 0.17°C, i.e., the temperature cannot change by more than 0.17°C from beginning to the end of any test run.*

6. *Tests are done at 120, 180, 230, and 275°C. Data is taken on sixteen RTDs, 1 to 16, 16 to 1, 1 to 16, 16 to 1. The four calibration points are used to determine a zero and a slope for the correction to temperature transmitters.*

N/A: Data Not Available

TABLE 23.3

Calibration Errors Without Ice Point

	Difference (°C)			
Tag	0°C	200°C	280°C	300°C
15A	0.09	0.013	0.002	0.006
15C	0.09	0.013	0.002	0.005
16A	0.09	0.014	0.003	0.004
16C	0.07	0.011	0.002	0.004
17A	0.08	0.013	0.002	0.005
17C	0.09	0.016	0.004	0.005

Above differences are between the fitting results of a four point calibration with and without the ice point.

TABLE 23.4

Narrow Range Calibration Errors

	Difference (°C)			
Tag	(0°C)	(200°C)	(280°C)	(300°C)

Calibration Points 300, 260, 200°C

Tag	(0°C)	(200°C)	(280°C)	(300°C)
15A	0.099	0.018	0.004	0.006
15C	0.072	0.017	0.003	0.006
16A	0.112	0.015	0.004	0.005
16C	0.121	0.013	0.004	0.004
17A	0.065	0.014	0.002	0.005
17C	0.129	0.040	0.004	0.005

Calibration Points 300, 260, 200, 160°C

Tag	(0°C)	(200°C)	(280°C)	(300°C)
15A	0.117	0.017	0.004	0.006
15C	0.109	0.016	0.003	0.006
16A	0.110	0.015	0.004	0.005
16C	0.107	0.013	0.004	0.004
17A	0.065	0.014	0.002	0.005
17C	0.071	0.016	0.004	0.004

Cross calibration is an effective method for verifying the consistency of a group of RTDs. The method does not account for common mode drift or any other systematic calibration problem unless one or more newly calibrated RTDs are included in the test. However, the fact that the drift of RTDs is usually random rather than systematic, as shown by the results of this project, justifies the use of cross calibration as a viable method even without including reference RTDs.

In lieu of cross calibration, a few plants perform calibration checks by thermodynamic calculations. In these approaches, the reactor coolant temperature is determined using steam pressure, feed water flow, etc. The data is compared with RTD indications and the differences are identified.

23.2 Accuracy of Cross Calibration Method

The uncertainties of the cross calibration method are discussed below and summarized in Table 23.5:

- Uncertainty of Measured Resistances. This depends on the accuracy and drift of the resistance measurement equipment. The best resistance measurement equipment available has accuracy and short term drift characteristics in the range of 0.005 to 0.01°C.

- Stability of the Process Temperature. Plant temperature fluctuations shall be accounted for in evaluating the overall accuracy of cross calibration. The cross calibration can be performed when the plant temperature is ramping up or ramping down at a constant rate. The advantage of this approach is better stability and the disadvantage is the error which would result if the ramp rate is not constant.

  Figure 23.1 shows in-plant cross calibration data for 16 RTDs. The data is shown before and after correcting for fluctuations. The correction involved implementing a curve smoothing procedure. Note that there is about 0.5°C of temperature fluctuation that has been reduced to about 0.1°C with the smoothing procedure. The data shown in Figure 23.1 is the result of four consecutive cross calibration repeats involving sixteen RTDs in each repeat.

  Table 23.6 shows stability data at 3 plateaus with and without correction for fluctuations. The data given is in terms of standard deviation of 64 data points representing four sets of cross calibration data on 16 RTDs. Note that the corrections improved the stability by about 0.06°C. This data along with results of similar measurements have shown that the uncertainties that result from plant temperature stability range from 0.03 to 0.1°C depending on whether or not the data is corrected for temperature stability.

- Uniformity. Because of incomplete mixing and differences in loop heat removal, the temperature of primary loops are not necessarily equal at isothermal conditions. We

TABLE 23.5

Cross Calibration Uncertainties

Uncertainty	Range (°C)
Resistance Measurement	0.005 - 0.01
Stability	0.030 - 0.10
Uniformity	0.065 - 0.21
RSS Error (°C)	0.07 - 0.23
Max Error (°C)	0.10 - 0.32

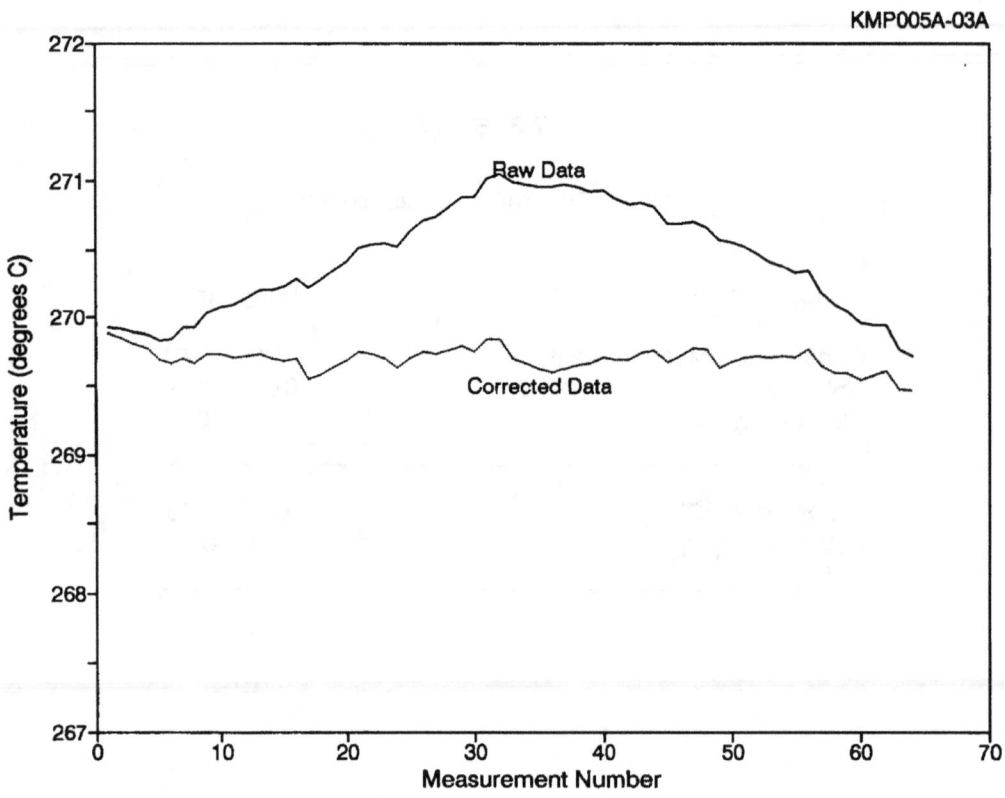

Figure 23-1.   Cross Calibration Data Before and After Correction
for Fluctuations.

TABLE 23.6

Temperature Stability Correction

Temperature Plateau (°C)	Run #	Stability (°C) Raw Data	Stability (°C) Fluctuation Removed
280°C	1	0.10	0.03
	2	0.08	0.03
220°C	1	0.12	0.02
	2	0.06	0.02
170°C	1	0.09	0.03
	2	0.11	0.03
Average		0.09	0.03

measured as much as 0.5°C difference between the hot leg and cold leg temperatures in the same loop of a two loop PWR at isothermal conditions and as much as 0.2°C across its core (Figure 23.2). These differences are referred to as uniformity error. The error can be identified and subtracted from the data to achieve better accuracy. The range of uniformity errors determined from cross calibration data in a two loop PWR was 0.065 to 0.21°C

To establish the validity of our cross calibration approach, we performed laboratory cross calibration tests on twenty newly calibrated RTDs at 300°C. The results are given in Table 23.7. Note that except for an unstable RTD (# 21) that was included intentionally here, almost all deviations are less than 0.05°C.

23.3 Combining the Cross Calibration Errors

The uncertainties given in Table 23.5 were estimated from the field test results obtained in this project. The lower limits for the uncertainty ranges in Table 23.5 are achievable if newly calibrated high precision resistance measurement equipment is used and the primary coolant temperature is very stable and uniform during the tests or corrections are made for stability and uniformity. To obtain the range of the total cross calibration error, we first assumed that all errors are random and combined them by calculating the root sum squared (RSS). This gave a range of 0.07 to 0.23°C. The errors were then assumed to be all systematic and their algebraic sum was found to have a range of 0.10 to 0.32°C. That is, the accuracy of cross calibration lies between 0.07 and 0.32°C depending on the accuracy of the resistance measurement equipment used in performing the cross calibration and the plant stability and uniformity during the cross calibration tests.

23.4 Cross Calibration Accuracy of Three-Wire RTDs

The uncertainties mentioned in Section 23.2 excluded the errors associated with any lead wire imbalances in three-wire RTDs. The error due to lead wire imbalance arises from differences that may exist between the resistances of the wires that extend from the sensing element to the resistance measurement equipment. Figure 23.3 shows a three-wire and a four-wire arrangement. In a four-wire arrangement, the lead wire resistances are completely compensated while in a three-wire arrangement, it is required that the resistance of wire 3 (R_3) be equal to the resistance of wire 1 (R_1) or wire 2 (R_2) depending on which wire is used as common wire in the three-wire bridge.

When using a precision digital multimeter to make measurements, the cross calibration procedure for three-wire RTDs involves three separate measurements to obtain the resistance of each RTD:

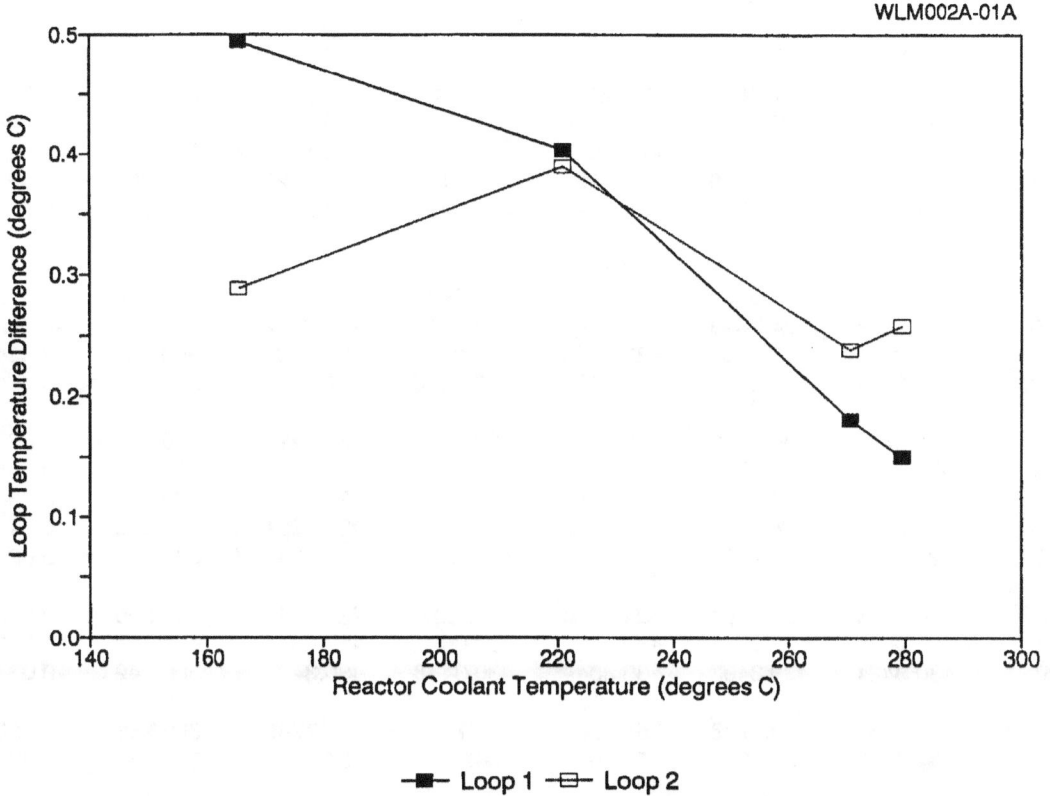

Figure 23-2. Difference Between the Hot Leg and
Cold Leg Temperatures in Each Loop
of a Two Loop PWR.

TABLE 23.7

Sample Results for Validation of
Cross Calibration of Four-Wire RTDs

XCAL0005.OUT   Plateau: 300°C

| Tag | Resistance Measurements (Ohms) | | | | Avg. Res. (Ohms) | Temp. (°C) | Dev. (°C) |
	Pass 1	Pass 2	Pass 3	Pass 4			
21	212.2644	212.2643	212.2741	212.2634	212.2666	*303.537	# 2.751
19	429.4928	429.4910	429.5060	429.4940	429.4960	300.759	-0.028
12A	429.1156	429.1152	429.1298	429.1134	429.1185	300.811	0.025
12C	429.0084	429.0088	429.0224	429.0102	429.0125	300.807	0.021
03	424.6416	424.6324	424.6384	424.6494	424.6405	300.785	-0.001
18	429.8286	429.8200	429.8228	429.8394	429.8277	300.819	0.033
15A	424.6800	424.6818	424.6862	424.6988	424.6867	300.774	-0.012
15C	424.2652	424.2698	424.2740	424.2818	424.2727	300.774	-0.012
13A	428.9974	428.9962	429.0018	429.0144	429.0025	300.762	-0.024
13C	428.9734	428.9702	428.9726	428.9840	428.9751	300.771	-0.015
9C	430.2376	430.2382	430.2478	430.2404	430.2410	300.787	0.001
9A	430.1488	430.1532	430.1658	430.1480	430.1540	300.780	-0.006
17A	424.3216	424.3148	424.3264	424.3152	424.3195	300.840	0.054
17C	424.2266	424.2230	424.2322	424.2186	424.2251	300.833	0.047
16A	424.7866	424.7882	424.7912	424.7800	424.7865	300.806	0.020
16C	424.5208	424.5294	424.5258	424.5228	424.5247	300.806	0.020
07	430.0636	430.0732	430.0620	430.0628	430.0654	300.724	-0.062
20	430.3424	430.3492	430.3344	430.3448	430.3427	300.749	-0.037
SPRT-1	54.7764	54.7778	54.7761	54.7767	54.7768	300.788	0.002
SPRT-2	54.7930	54.7954	54.7930	54.7935	54.7937	300.762	-0.024

Average Temperature: 300.786

Not used in average
 # Deviation limit exceeded

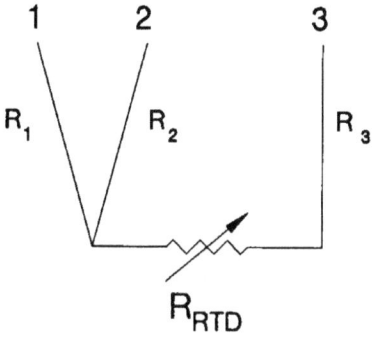

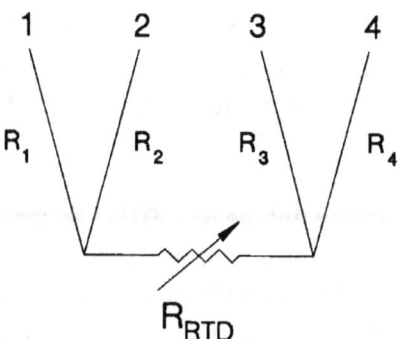

Figure 23-3. Three-Wire and Four-Wire RTD Configurations.

$Measurement\ 1:\ R_{13} = R_1 + R_{RTD} + R_3$ (23.1)

$Measurement\ 2:\ R_{23} = R_2 + R_{RTD} + R_3$ (23.2)

$Measurement\ 3:\ R_{12} = R_1 + R_2$ (23.3)

For accurate results, it is required that:

$$R_3 = \frac{R_1 + R_2}{2}$$ (23.4)

Then

$$R_{RTD} = \frac{Measurement\ 1\ +\ Measurement\ 2}{2} - Measurement\ 3$$ (23.5)

The resistance of wire 3 cannot be measured remotely. Therefore, it is not possible to determine the exact impact of the lead wire imbalances. What can be done is to identify the values of R_1 and R_2 and estimate the uncertainties due to lead wire imbalances based on the differences between R_1 and R_2. This was done for a set of cross calibrations performed in a PWR which included sixteen three-wire RTDs. The results are shown in Table 23.8. The average difference between the resistances of the two wires is about 0.07 ohms. This corresponds to an uncertainty of about 0.09°C for a 200 ohm RTD.

The total error in cross calibration of a three-wire RTD has the same components as those shown in Table 23.5 except for the uncertainties associated with lead wire imbalances. Furthermore, in validating the cross calibration method, we performed laboratory cross calibration on twenty newly calibrated RTDs in an oil bath at 300°C with four-wire and three-wire configurations. The three-wire tests were performed with the four-wire RTDs used in three-wire configurations. Sample results for four-wire cross calibrations were shown in Table 23.7. Sample results for 3 wire RTDs are shown in Table 23.9. Note that the deviations are larger than those seen in Table 23.7.

TABLE 23.8

Lead Wire Imbalance at 280°C Plateau

	Resistance (ohm)		
Tag	R_1	R_2	ΔR ($R_1 - R_2$)
1	2.228	2.476	0.248
2	2.968	2.955	0.013
3	2.257	2.303	0.047
4	2.310	2.417	0.108
5	2.302	2.311	0.009
6	2.239	2.284	0.045
7	2.916	3.018	0.102
8	2.243	2.251	0.008
9	2.567	2.671	0.095
10	2.530	2.540	0.010
11	2.304	2.373	0.069
12	2.088	2.072	0.016
13	2.552	2.662	0.070
14	2.248	2.218	0.030
15	2.697	2.636	0.061
16	2.336	2.098	0.238

Average of Δ (Ω)	0.073
Temperature Error (°C)	0.091

TABLE 23.9

Sample Results of Validation of
Cross Calibration of Three-Wire RTDs

XCAL0016.OUT   Plateau: 300°C

Tag	Pass 1	Pass 2	Pass 3	Pass 4	Avg. Res. (Ohms)	Temp (°C)	Dev. (°C)
21	212.2397	212.2376	212.2349	212.2344	212.2366	*303.452	# 2.908
19	429.3999	429.4064	429.3937	429.3864	429.3966	300.622	0.078
12A	428.8234	428.8349	428.8227	428.8128	428.8234	300.403	-0.141
12C	428.7745	428.7909	428.7758	428.7656	428.7767	300.481	-0.063
03	424.7646	424.7887	424.7765	424.7661	424.7740	*300.973	# 0.429
18	429.6882	429.7120	429.7032	429.6962	429.6999	300.643	0.099
15A	424.5699	424.5908	424.5827	424.5752	424.5797	300.623	0.079
15C	424.1568	424.1741	424.1683	424.1659	424.1663	300.625	0.081
13A	428.7102	428.7284	428.7178	428.7214	428.7195	300.371	#-0.173
13C	428.6318	428.6460	428.6365	428.6451	428.6399	300.308	#-0.236
9C	430.0284	430.0315	430.0301	430.0450	430.0338	300.500	-0.044
9A	429.9458	429.9449	429.9430	429.9646	429.9496	300.497	-0.047
17A	424.1572	424.1518	424.1428	424.1502	424.1505	300.602	0.059
17C	424.0595	424.0545	424.0462	424.0589	424.0548	300.593	0.049
16A	424.6559	424.6485	424.6553	424.6712	424.6577	300.625	0.081
16C	424.3978	424.3937	424.4001	424.4085	424.4000	300.631	0.087
07	429.8613	429.8435	429.8580	429.8836	429.8616	300.445	-0.099
20	430.2722	430.2506	430.2682	430.2763	430.2668	300.645	0.101
SPRT-1	54.7627	54.7614	54.7617	54.7634	54.7623	300.632	0.088
SPRT-2	54.8013	54.7996	54.8012	54.8046	54.8017	*300.848	# 0.304

Average Temperature: 300.544

Not used in average
 # Deviation limit exceeded

- 154 -

23.5 Plateau Method Versus Ramp Method

Cross calibration may be done with the plant at isothermal condition as has been described in the above sections, or when the plant temperature is undergoing a ramp change. The ramp test is accurate only if the ramp rate is constant. The RTDs are scanned from first to last and then from last to first and the results are averaged in the same manner as the plateau method.

The validity of the ramp method was established by performing cross calibration in an oil bath with twenty newly calibrated RTDs. The results are shown in Table 23.10 for a case when the bath temperature was ramping down from 300°C to room temperature at a rate of about 60°C/hr. The results of repeated ramp tests performed in this project were usually close to those of plateau when the ramp rates were relatively slow and constant.

The results in Table 23.10 are for four-wire configurations of the same RTDs as in Table 23.7.

TABLE 23.10

Sample Results of Validation of
Cross Calibration with Temperature Ramp

XCAL0018.OUT   Plateau: Ramp

Tag	Resistance Measurements (Ohms)				Avg. Res. (Ohms)	Temp (°C)	Dev. (°C)
	Pass 1	Pass 2	Pass 3	Pass 4			
21	212.1559	209.3864	209.3155	206.6431	209.3752	*295.338	#2.631
19	429.1486	423.7962	423.3428	418.2018	423.6224	292.681	-0.026
12A	428.7144	423.6962	422.9530	418.1374	423.3752	292.885	#0.177
12C	428.4792	423.7342	422.7188	418.1828	423.2788	292.892	#0.184
03	423.8728	419.4584	418.1750	413.9560	418.8656	292.686	-0.022
18	428.8116	424.5744	422.9724	418.9354	423.8235	292.575	-0.132
15A	423.7890	419.9490	418.1166	414.4610	419.0789	292.900	#0.193
15C	423.2386	419.6854	417.5794	414.1920	418.6739	292.905	#0.197
13A	427.7112	424.3816	421.9590	418.7786	423.2076	292.766	0.058
13C	427.5354	424.4894	421.7880	418.8812	423.1735	292.765	0.057
9C	428.5696	425.8194	422.8422	420.2084	424.3599	292.664	-0.043
9A	428.3434	425.8806	422.6212	420.2622	424.2769	292.663	-0.045
17A	422.3640	420.2154	416.7224	414.6750	418.4942	292.655	-0.052
17C	422.1288	420.2594	416.5016	414.7212	418.4028	292.646	-0.061
16A	422.5072	420.9188	416.8842	415.3698	418.9200	292.567	-0.141
16C	422.1058	420.8016	416.4910	415.2438	418.6606	292.565	-0.143
07	427.4872	426.4386	421.6986	420.6922	424.0792	292.526	#-0.182
00	427.0042	420.9100	421.9240	421.1984	424.4262	292.617	-0.090
SPRT-1	54.4276	54.3723	53.6961	53.6425	54.0346	292.793	0.085
SPRT-2	54.4188	54.4000	53.6885	53.6699	54.0443	292.692	-0.016

Average Temperature: 292.707

Not used in average
 # Deviation limit exceeded

- 156 -

24. RESPONSE TIME TESTING EXPERIENCE

The speed of response of primary system RTDs is crucial to safety in case of a sudden change in reactor coolant temperature. In the safety analysis of nuclear power plants, a limit is usually specified for the response time of safety system RTDs. Since RTD response time degradation occurs, periodic response time measurements are made to ensure that the safe limits are not exceeded while the plant is operating. These measurements are independent of RTD calibration and are therefore performed in addition to calibration.

The need for RTD response time measurements was recognized in the mid 1970's when the first draft of Regulatory Guide 1.118 was issued by the NRC recommending periodic sensor response time testing in nuclear power plants. This recommendation led to the development of the Loop Current Step Response method which has been in routine use in the nuclear industry for over ten years. Normally, ten years of experience with response time measurements performed in numerous plants should provide a reliable database of aging effects on response time. Unfortunately, this is not the case because in the last five years, most of the RTDs for which response time history was accumulated have been replaced with new RTDs for environmental and seismic qualification concerns and product improvement reasons.

The first major problem with response time of nuclear plant RTDs was identified in the early 1980's. This was with well-type RTDs which used a thermal coupling compound called "Never-Seez" in their thermowells. This compound helps reduce the response time of the RTD when it is fresh. However, at plant operating conditions, Never-Seez degrades and causes the response time to increase. Never-Seez is no longer used for this purpose except in a few plants.

Without a thermal compound, there is a strong sensitivity of response time to seating of the RTD into its thermowell. Laboratory test results have shown that a fraction of a millimeter of air gap in the RTD/thermowell interface can cause the response time to increases significantly. Table 24.1 shows the increases in response time of a tapered-tip RTD as it was removed out of the thermowell a few hundredths of a millimeter at a time. The experiment was conducted with the test fixture illustrated in Figure 24.1. The plates and springs shown in Figure 24.1 were installed to permit systematic displacement of the RTD from the tip of its thermowell. A feeler gage as shown in the figure was used to measure the axial displacement of the RTD. This experiment simulates the conditions that may occur during in-plant use of the RTD. Note that the RTD time constant increased by more than 60 percent with an air gap of about 0.04 mm (less than two thousands of an inch). Additional research results on the response time of RTDs are found in References 7 and 8.

TABLE 24.1

Response Time Changes in Well-Type RTDs

Change in Gap Size (mm)	Time Constant (Seconds)
0	6.3
0.006	7.1
0.010	8.0
0.016	8.9
0.022	9.5
0.035	10.5

mm = millimeter

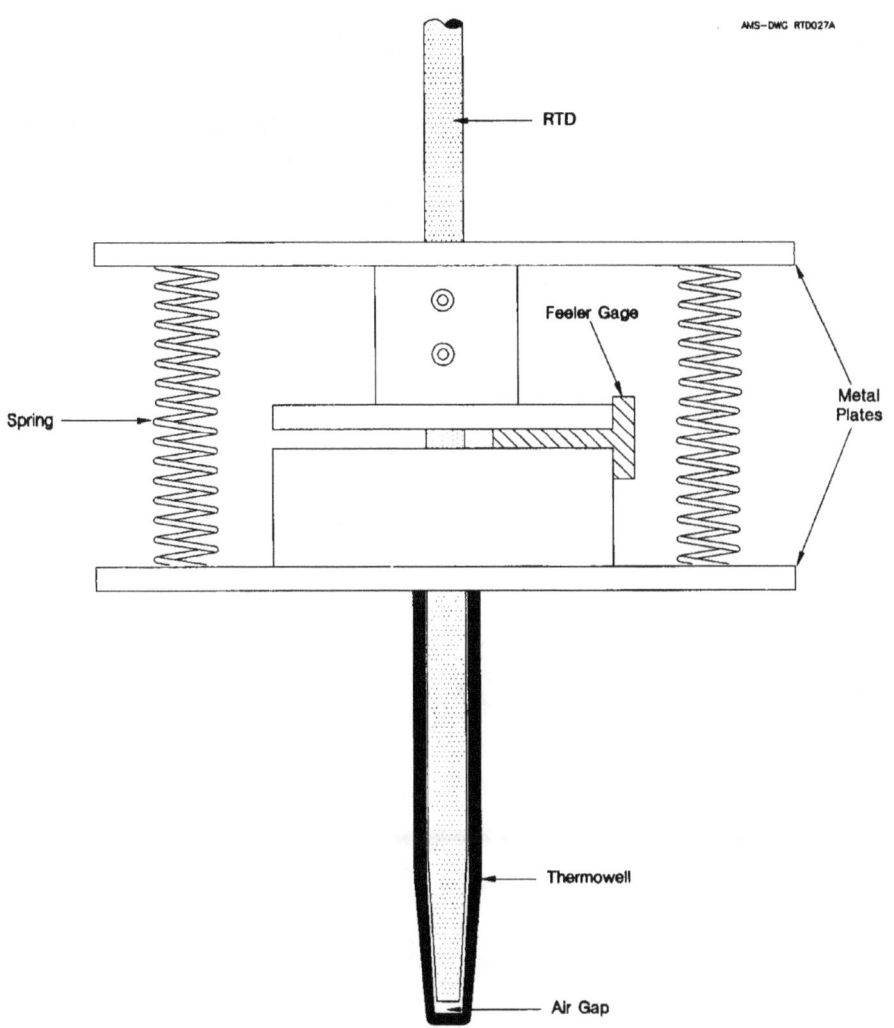

AMS–DWG RTD027A

RTD

Feeler Gage

Metal
Plates

Spring

Thermowell

Air Gap

Figure 24-1. Test Fixture for RTD Displacement Test.

The response times of well-type RTDs, excluding those with obvious installation problems, are in the range of 3 to 8 seconds and the average is about 5 seconds. This conclusion is based on a review of in-service time constants of about two hundred RTDs representing the three most popular well-type RTDs used in the nuclear industry. We must point out that two identical RTDs from the same manufacturer can have response times of as little as 3 seconds or as large as 8 seconds when installed in different thermowells.

For direct immersion RTDs, the range of response time depends on the design. Direct immersion RTDs are available with response times from less than 1 second to over 5 seconds. The most widely used direct immersion RTD is one that has a response time of 2 to 4 seconds with the average of about 3 seconds. This conclusion is based on a review of in-plant test results for about one hundred direct immersion RTDs in operating PWRs.

25. ACCEPTABLE TEST METHODS

Examples of acceptable test methods for calibration and response time testing of nuclear plant RTDs are discussed here. This discussion includes the methods for pre-installation laboratory testing as well as in-plant testing at normal operating conditions.

25.1 Testing for RTD Accuracy

An acceptable method for verifying RTD accuracy is a full laboratory calibration using a procedure similar to that described in Section 21. Alternative methods and their acceptability are discussed below.

A check of ice point resistance of an RTD in lieu of a full calibration is not usually effective in verifying accurate performance at high temperatures. Table 25.1 shows the results of calibrations of nine RTDs that have negligible ice point deviations but large errors at higher temperatures. The results are shown in terms of the differences at ice point versus the maximum differences in the range of 0 to 300°C seen in the results of four consecutive calibrations.

Calibration using an ice bath whose temperature is not measured with an SPRT is not acceptable unless the ice bath is properly made of distilled water and distilled ice. Boiling water cannot be used for calibration unless its temperature is measured and adequate data is taken and averaged to overcome any temperature fluctuations due to boiling. The uniformity of the boiling water bath must be accounted for in performing accurate calibrations.

The differences between a precise calibration and four casual calibrations are shown in Table 25.2 for five RTDs. This experiment was conducted to verify the benefit of a precise calibration. The results given are the maximum differences in the range of 0 to 300°C between a precise calibration and casual calibrations involving four or five calibration points. Note that errors of about 0.2 to 2.2°C are encountered depending on how well the calibrations are performed. For precise calibrations, the errors are in the range of 0.02 to 0.2°C, an order of magnitude better than the errors of casual calibrations.

For an acceptable calibration, the accuracy of the calibration equipment and the calibration process must be determined and documented. The calibration tables should list the calibration points and give temperatures to three significant digits and resistances to four significant digits. Calibration tables should not be extrapolated to more than 20 percent above the highest temperature at which the RTD is calibrated.

TABLE 25.1

Calibration Shift Versus Ice Point Shift

Tag	Initial to Cal. 1		Cal. 1 to Cal. 2		Cal. 2 to Cal. 3	
	Ice Point	Max. Shift	Ice Point	Max. Shift	Ice Point	Max. Shift
57	0.001	0.02	0.005	0.01	0.001	0.00
62	0.000	0.05	0.005	0.01	0.001	0.00
63	0.003	0.60	0.007	0.04	0.001	0.08
60	0.004	0.24	0.000	0.02	0.004	0.08
59	0.000	0.33	0.001	0.16	0.004	0.17
58	0.001	0.53	0.001	0.02	0.003	0.04
61	0.006	0.13	0.003	0.05	0.004	0.04
56	0.003	0.01	0.002	0.01	0.003	0.01
15	0.003	0.01	0.001	0.01	0.001	0.01

Above data are temperature shifts in °C. The ice point shift is the difference at 0°C between the results of two calibrations. Max. shift is the maximum difference in the range of 0 to 300°C between each pair of consecutive calibrations.

TABLE 25.2

Errors in Casual Calibration of RTDs

Tag	Error (°C)			
	A	B	C	D
35	2.5	1.7	0.8	0.4
43	2.0	1.4	0.4	0.2
44	2.2	1.5	0.6	0.2
74	2.2	1.4	0.5	0.3
95	1.8	1.2	0.1	0.3

Calibration Points

A. *Four point calibration. This calibration was conducted in the following media: boiling water bath (assumed to be 100°C), ice bath (assumed to be 0°C), room temperature bath (measured with a regular thermometer), and an oil bath at 250°C (measured with a regular thermometer).*

B. *Five point calibration. this calibration was conducted in the following media: boiling water bath (assumed to be 100°C), ice bath (assumed to be 0°C), room temperature bath (measured with a regular thermometer), oil bath set at 80°C, and an oil bath set at 250°C (both measured with a regular thermometer).*

C. *Five point calibration. This calibration was conducted in the following media: boiling water bath (measured with a regular thermometer), ice bath (assumed to be 0°C), room temperature water bath (measured by a regular thermometer), oil bath set at 80°C, and an oil bath set at 250°C (both measured with a regular thermometer).*

D. *Five point calibration. This calibration was conducted in the following media: boiling water bath (measured with an SPRT), ice bath (assumed as 0°C), room temperature bath (measured with an SPRT), oil bath at 80°C (measured with an SPRT), and an oil bath at 250°C (measured with an SPRT).*

Installed RTDs may be calibrated using the cross calibration method. At least one independently calibrated and carefully installed RTD must be included in the first series of in-plant cross calibrations to verify that there is no bias in the RTDs. A major disadvantage of the cross calibration method is that it will not detect unidirectional drift or a systematic problem in the RTDs unless a few newly calibrated RTDs are included in each set of cross calibration.

The Johnson noise technique and cross calibration tests that include thermocouples are not acceptable. The Johnson noise technique cannot resolve small enough differences when several hundred feet of extension wires are involved. Thermocouples should not be included in any cross calibration of RTDs as thermocouples are not generally as accurate and reliable as RTDs and cannot be recalibrated once they are used in the plant.

RTD removal from the plant for recalibration in a laboratory is marginally acceptable. On one hand, this method is unquestionable because the best way to ensure accuracy is to perform a laboratory calibration and on the other hand, the method is undesirable because of the possibility of damage to the RTDs and calibration shifts during removal and installation.

25.2 Response Time Testing

The acceptable method for measurement of response time of an RTD is the Loop Current Step Response (LCSR) test. The test must be performed at or near normal operating conditions to yield the in-service time constant of the RTD. Since the validity of the LCSR test depends on certain assumptions, the LCSR testability of each RTD design must be established by laboratory testing. The laboratory validation tests require performing plunge tests and LCSR tests in a reference condition to verify that the LCSR method can provide results within 10 percent of the actual step response of the RTD.

Self heating tests can be used to supplement the LCSR results. This test does not provide a time constant, but is useful for detecting gross changes in RTD response characteristics. The test involves measuring the steady state increase in RTD resistance as a function of applied power. This results in the self heating index of the RTD in terms of ohms/watt. This index is proportional to response time and its sensitivity depends on the RTD design and heat transfer characteristics. For some RTDs, the sensitivity of the self heating index is very good and for others, the sensitivity is so small that it can only detect very large changes in response time.

RTD removal for response time testing in a reference condition in a laboratory is not acceptable. This is because the response time of an RTD depends on installation and process conditions that can not be simulated. This is especially true with well-type RTDs for which there is very little relationship between the RTD time constant in the thermowell used in laboratory tests and the thermowell in the plant.

26. TESTING INTERVALS AND REPLACEMENT SCHEDULES

The current industry practice for verifying adequate RTD accuracy and response time is to perform on-line cross calibration and response time testing at least once every fuel cycle. In light of the results generated in this project, this practice is reasonable unless there are plant specific problems requiring more frequent testing or the RTDs are suspected of deficiencies in design, fabrication, or installation. For example, in one plant, a small margin between the required response time and the nominal response time of primary coolant RTDs, in addition to a history of response time problems due to degradation of a thermal compound used in the thermowell, required periodic response time testing to be performed once every one or two months.

The data available on drift and response time degradation of RTDs including those provided in this report are so random that a reliable rate of change for either calibration or response time of RTDs can not be established. Therefore, RTD replacement schedules should be based on performance problems identified during the periodic in-plant tests. For example, an RTD that has consistently shown measurable monotonic drift in either positive or negative directions should be replaced. Any RTD that has suffered a shift of more than 1°C should be replaced. Any major change or consistent increases in response time of well-type RTDs should be followed by an attempt to clean and reseat the RTD in the cleaned thermowell. This may or may not resolve the problem. If not, the RTD and sometimes even the thermowell may have to be replaced. Any direct immersion RTD that has been found to have an unacceptable response time should be replaced as there is no other way to restore the response time of direct immersion RTDs.

A good time for performing in-plant cross calibration or response time testing is near the end of a fuel cycle prior to a refueling outage. This allows for any RTD problem to be resolved during the outage. If "as found" and "as left" calibration data is required, a few RTDs should be removed at each refueling outage, recalibrated, and the results used along with the cross calibration data to determine the "as found" status of the plant RTDs.

Those RTDs that consistently pass response time and calibration testing can be kept and used in the plant for their design life as specified by the manufacturer. Typical design life of nuclear grade RTDs is 10 to 40 years depending on the conditions of use.

27. SEARCH OF LER AND NPRDS DATABASES

The LER and NPRDS databases were searched for specific reports of failures of nuclear plant RTDs. The results are summarized here. The details are given in Appendix B for the LER search and Appendix C for the NPRDS search. The NPRDS search was documented in a comprehensive report submitted to the NRC in December 1989[9].

The LER database had approximately 92 specific failure reports on RTDs during the period of 1970 to 1988. These failures are categorized below.

Description	No. of Failures	Percent of Total
Circuit Defects	32	35
Drift	15	16
Moisture	15	16
Connection/Wiring	15	16
Response Time	9	10
Other	6	7

The NPRDs database had 318 reports of failure during the 1974 to 1988 period. These are summarized in Table 27.1. Note that both the LER and NPRDS data bases point to circuit defect as the dominant cause of failures. Figure 27.1 illustrates the approximate distribution of the NPRDS reports. Note that the total number of reports referred to in Table 27.1 adds up to more than 318 because of multiple categories used for a few of the failures.

The small number of failure reports in the LER and NPRDS databases are probably due to the fact that prior to 1980, primary coolant RTDs were not tested alone to identify any response time or calibration problems. They were sometimes included in the tests of a temperature channel or were not tested at all.

TABLE 27.1

Summary of Search of NPRDS Database
for Failures of Nuclear Plant RTDs

Cause Description	No.of Reports
Abnormal Stress[1]	15
Aging/Cycling Fatigue[1]	26
Burned/Burned Out[2]	5
Circuit Defective[2]	68
Connection Defective/Loose Parts[2]	31
Contacts Burned/Pitted/Corroded[1]	1
Corrosion[1]	11
Dirty[1]	7
Foreign/Incorrect Material[3]	2
Foreign/Wrong Part[3]	1
Incorrect Action[3]	12
Incorrect Procedure[3]	1
Insulation Breakdown[1]	11
Material Defect[2]	7
Mechanical Damage/Binding[2]	9
Normal/Abnormal Wear[1]	41
Open Circuit[2]	54
Previous Repair/Installation Status[3]	13
Out of Calibration[1]	36
Out of Mechanical Adjustment[1]	3
Particulate Contamination[1]	2
Setpoint Drift[1]	6
Short/Grounded[2]	34

Combined Categories

[1]Potential Age-Related	159
[2]RTD or Circuit Defect	208
[3]Personnel Related	29

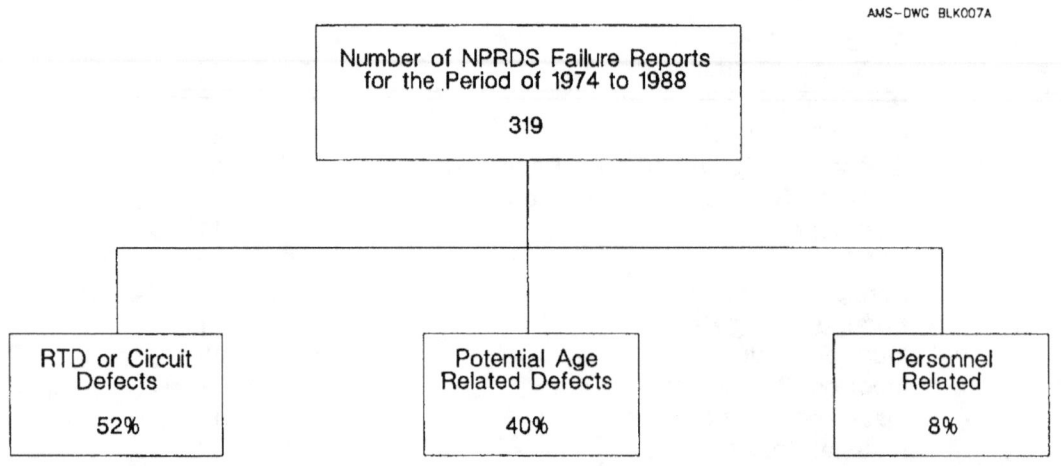

Figure 27-1. Distribution of NPRDS Failure Reports.

28. RTD BY-PASS MANIFOLD ELIMINATION

By-pass loops are used in Westinghouse PWRs to overcome temperature stratification problems and provide for easier replacement of RTDs. The disadvantage of this system is that it adds about 300 feet of piping with associated valves and other hardware to the primary system of the plant. Experience has shown that the by-pass loops are a source of radiation due to radioactive material that becomes trapped in the system. These problems have recently motivated the nuclear industry to remove the by-pass loops and install well-type RTDs directly into the reactor coolant pipes. To date, fourteen nuclear power plants have successfully implemented RTD by-pass manifold elimination projects.

Figure 28.1 shows a simplified schematic of a PWR with and without the by-pass manifolds. The system with the by-pass loops uses direct immersion RTDs with response times of less than 3.0 seconds. The fast response compensates for the time lag required for the primary coolant to reach the by-pass manifold in which the RTDs are installed. This is referred to as the transport time delay.

Table 28.1 shows the system response times in a PWR before and after RTD by-pass removal[10]. The RTD response time of 4.75 seconds includes a 10 percent allowance for accuracy of the Loop Current Step Response method. Note that the slow response of well-type RTDs is overcome by eliminating the transport time delay. The 0.25 seconds shown is for the transport delay inside the hot leg scoops in which the RTDs are located. There are three scoops 120° apart around each hot leg pipe in Westinghouse plants (Figure 28.2). In the original system, scoops were used to sample the reactor water from three locations in the pipe to provide a well-mixed water sample for an average temperature measurement. In the new system, this is accomplished by averaging the reading of the individual RTDs in the plant primary system.

The response time of well-type RTDs is extremely sensitive to installation in a thermowell. To avoid RTD response time problems after by-pass manifold elimination, most plants implement a set of LCSR tests performed at cold shutdown to identify the outliers. The outliers are replaced, reseated, or reinstalled and the tests are repeated to verify that the problem is resolved. Table 28.2 shows examples of typical results. Note that these response time are from tests at cold shutdown conditions and do not, therefore, represent the in-service response times of the RTDs. The variations in time constants shown in Table 28.2 are due to natural circulation or RHR induced flow that exists at cold shutdown around some RTDs and not the others. The RHR (Residual Heat Removal) system provides shutdown cooling by removing water from one hot leg and discharging it to one or two cold legs after it passes through a heat exchanger.

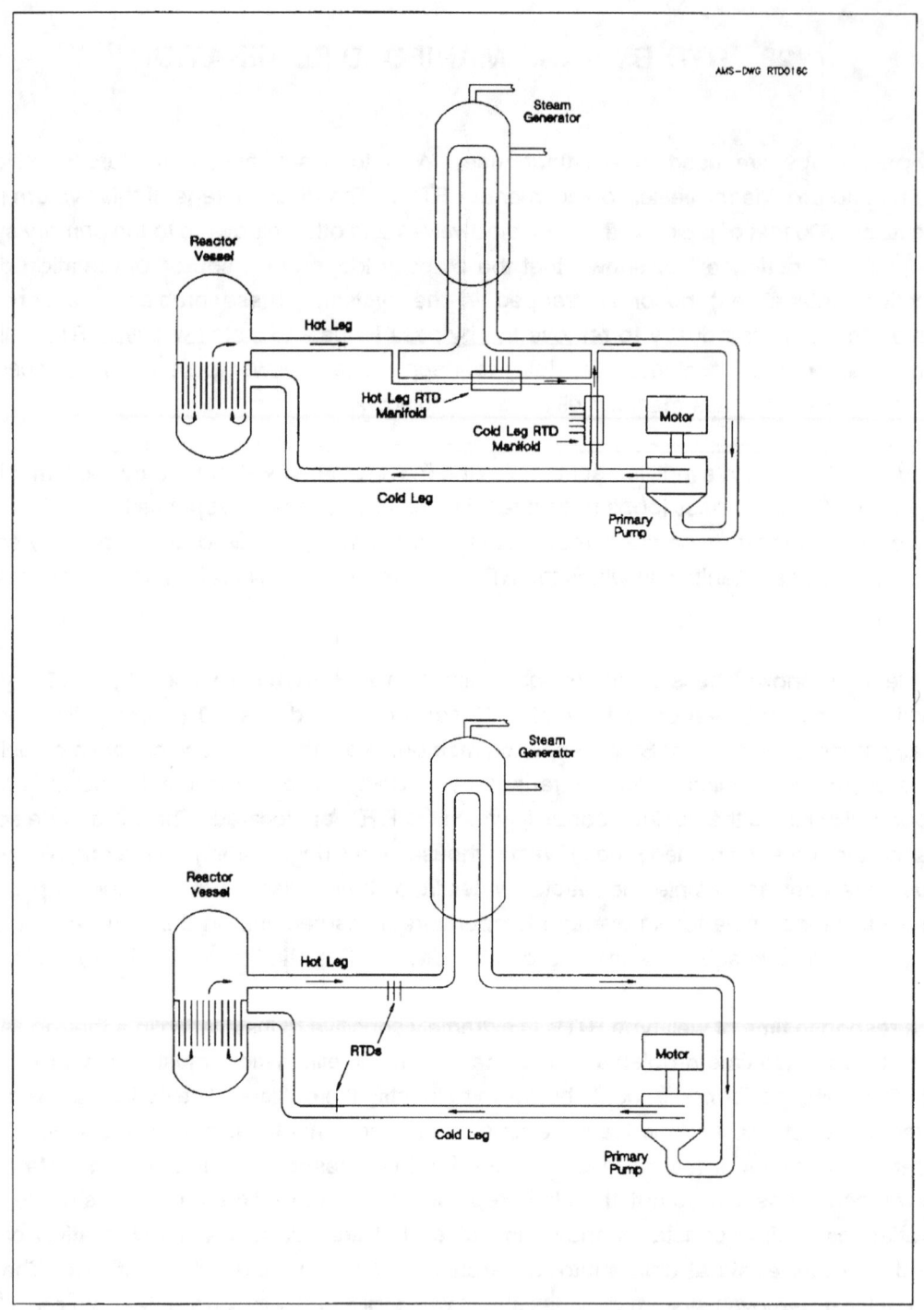

Figure 28-1.   Simplified Schematic of a PWR Primary System
With and Without RTD By-Pass Loops.

TABLE 28.1

System Response Time Before and After
Removal of RTD By-Pass Manifolds

Response Component	Time Constant (sec)	
	W/By-Pass	W/O By-Pass
RTD	3.0	4.75
Electronic	1.0	1.0
Transport or Mixing	2.0	0.25
Total	6.0	6.0

W/By-Pass: *Before removal of the manifolds*
W/O By-Pass: *After removal of the manifolds*

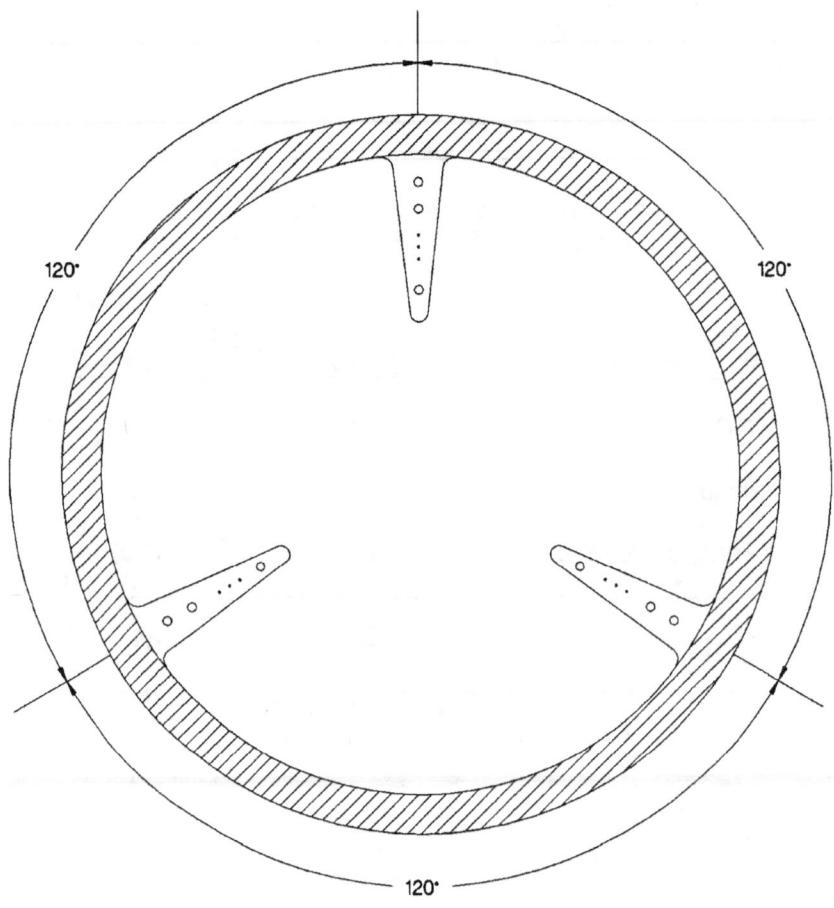

Figure 28-2.   Hot Leg Scoops in Primary Coolant
System of Westinghouse Plants.

TABLE 28.2

Examples of Response Problems Identified
and Resolved At Cold Shutdown

Time Constant (sec)		
As Found	As Left	Corrective Action
11.6	4.7	Cleaned Thermowell
22.5	7.5	Cleaned Thermowell
14.5	5.4	Installed New RTD
24.0	7.8	Installed New RTD
9.0	5.0	Reseated RTD
18.0	14.0	Reseated RTD
14.7	5.9	Cleaned Thermowell
15.5	6.2	Cleaned Thermowell

Since the response times of RTDs are affected by the process temperature and flow conditions, the results at cold shutdown cannot be used to relate to the in-service response times. The in-service response times of the RTDs must be obtained by LCSR testing at hot standby (or Mode 3 as it is referred to in some plants) conditions with the reactor coolant temperature above 260°C and a flow of at least 50 percent. The LCSR test is most conveniently performed at normal operating conditions, as opposed to hot standby conditions. However, after a by-pass manifold elimination project, most plants must verify satisfactory response time results before the plant can resume normal power operation.

In addition to response time, the accuracy of the measured temperature after by-pass removal must be verified to be equal to or better than that of the original system. This requires a knowledge of uncertainties involved in temperature measurement with the original system and the new RTDs. The challenge is in demonstrating the temperature stratification problem is effectively addressed without the by-pass manifolds.

29. INTERNATIONAL TEMPERATURE SCALE OF 1990

A new temperature scale called the International Temperature Scale of 1990 (ITS-90) has been adopted effective January 1, 1990 and is now being implemented in the United States by the National Institute of Standards and Technology (NIST). The ITS-90 replaces the International Practical Temperature Scale of 1968 (IPTS-68). The main differences between IPTS-68 and ITS-90 are:

- The lower range of IPTS-68 was -259.34°C. The new scale extends down to -272.50°C.

- The IPTS-68 was inaccurate in reproducing thermodynamic temperatures especially in the range of 630.74 to 1337.58°C. These inaccuracies have been improved by replacing type S thermocouples with SPRTs as the standard interpolation instrument for most of this range. Type S thermocouples (platinum/rhodium) were used in the IPTS-68 as the standard interpolation instrument for the range of 630.74 to 1337.58°C. In ITS-90, SPRTs are used from 630.74 to 961.78°C and radiation pyrometers are used for higher temperatures. Since SPRTs can be made to provide much better accuracies than thermocouples, the inaccuracies of IPTS-68 are reduced in ITS-90.

The difference between ITS-90 and IPTS-68 is shown in Figure 29.1[11]. The impact of the change in measurement of reactor coolant temperatures in PWRs is a correction of up to about 0.05°C in the range of 0 to 400°C. The correction curve is shown in Figure 29.2. The procedure for implementing this correction is given in Appendix D. It is apparent from the data shown in Figure 29.2 that the current temperature measurements in the range of 0 to 400°C are conservative, i.e., larger than the thermodynamic temperatures. As such, the new temperature scale does not result in a safety concern in measurement of reactor coolant temperatures in PWRs.

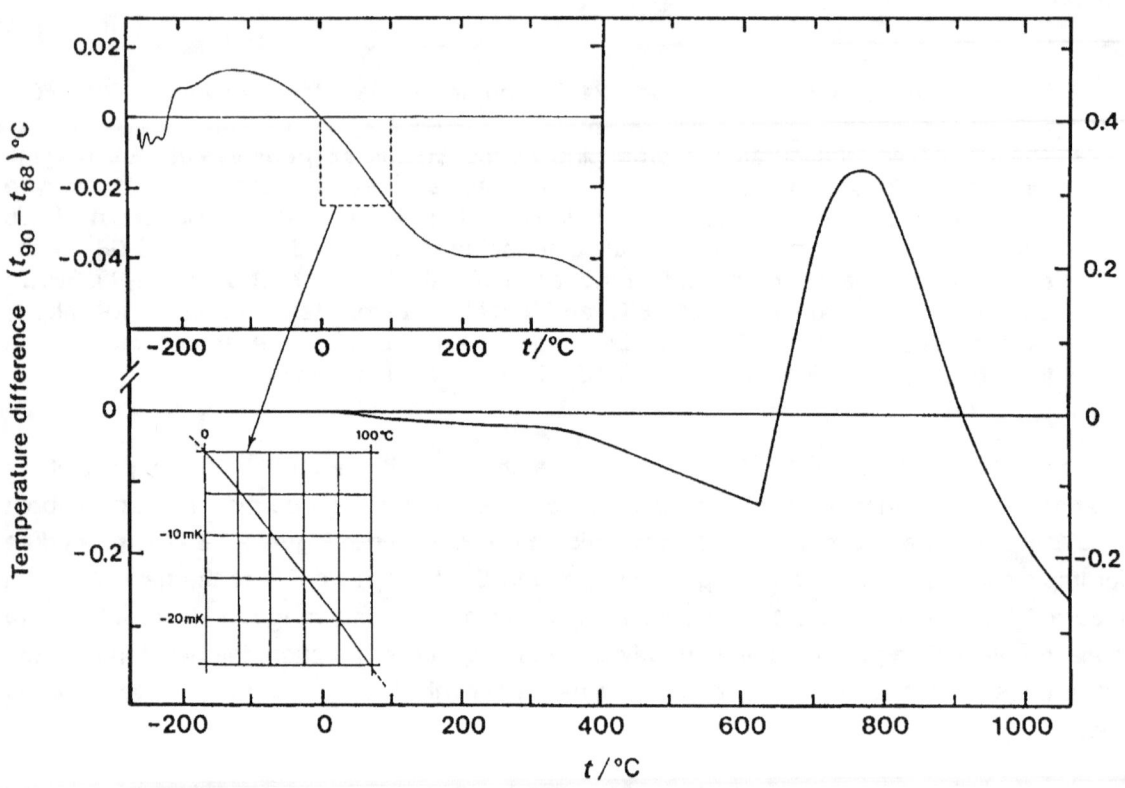

Figure 29-1. Difference Between ITS-90 and IPTS-68 Temperatures. (Figure reproduced from Reference 11)

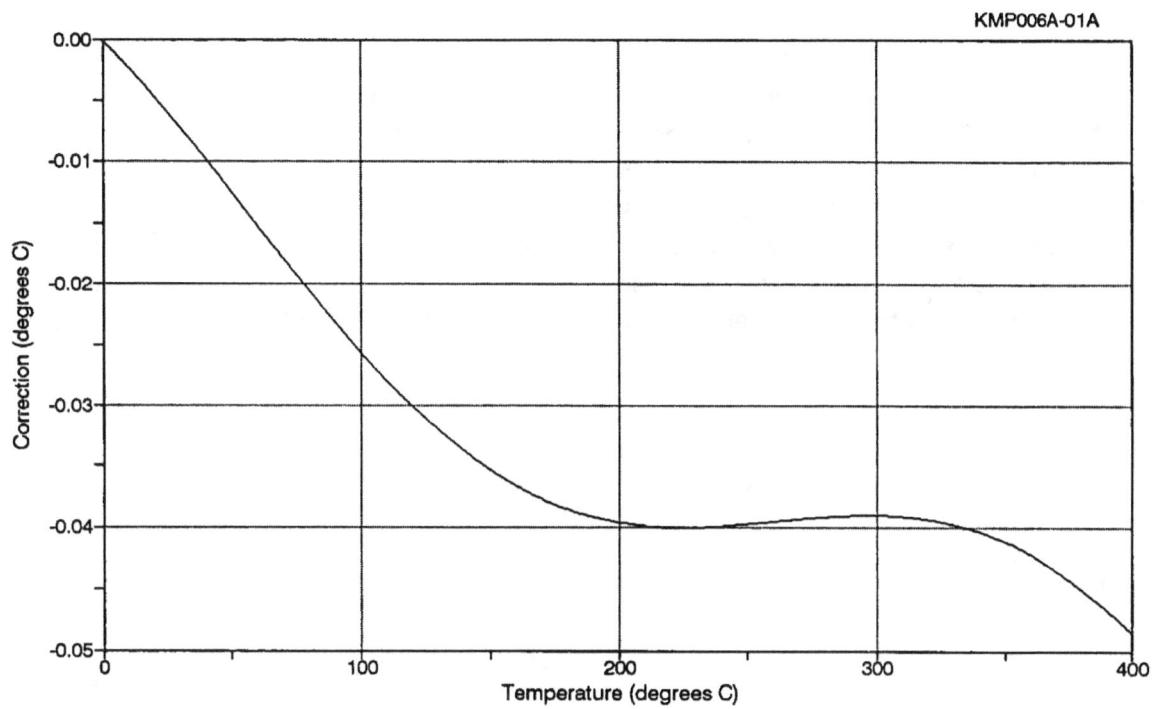

Figure 29-2.   Impact of ITS-90 on Reactor Coolant
                Temperatures in PWRs.

30. OTHER BENEFITS OF PRIMARY COOLANT RTDs

In addition to providing a measurement of the reactor coolant temperature, primary system RTDs can be used effectively at cold shutdown or in accident conditions to determine if the primary system is solid and to distinguish between stagnant and flowing water conditions. Furthermore, gross changes in primary coolant flow can be detected with RTDs.

As in the Loop Current Step Response test, the procedure is to apply a small electric current to the extension wires across the RTD element. The test can be done remotely from outside the containment. The applied current causes a temperature transient in the RTD, the decay rate of which would depend on the heat transfer conditions around the RTD. If the RTD is in air, the transient will approach steady state at a slow rate compared to when the RTD is in water. Figure 30.1 shows the transient response of an RTD in stagnant air, in air flowing at 15 meter/sec, in stagnant water, and in water flowing at 1 meter per second. It is clear that the test signal has strong diagnostics capability for the reactor coolant flow conditions around the RTD.

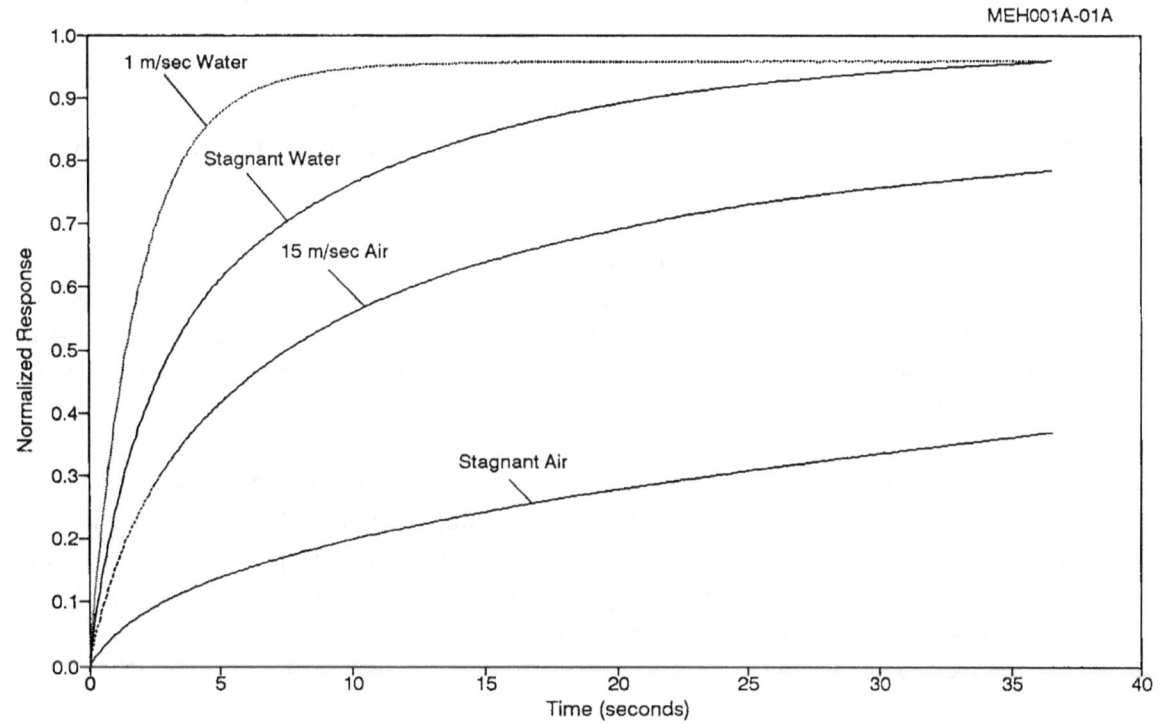

Figure 30-1.   Internal Step Response of an RTD in Various
Heat Transfer Environments.

31. CONCLUSIONS AND RECOMMENDATIONS

The calibration and response time of RTDs are affected by aging even at normal operating conditions. However, the aging is manageable by periodic tests performed at each refueling interval. These tests can be performed on-line using methods that have been developed and validated for use in nuclear power plants.

Prior to installation into a plant, each RTD should be calibrated and response time tested in a laboratory using equipment with valid calibration traceable to NIST. Any RTD that exhibits instability during calibration or has a long response time (in a matching thermowell if it is a well-type RTD) should not be installed in the plant. If an RTD has been in storage for more than two years, it should be recalibrated before it is installed in the plant. The same argument applies to RTDs that have been inactive such as those installed in a non-operating plant for a period of more than two years. The stability of these RTDs may be improved if they are first annealed and then calibrated.

The limit for the initial accuracy of nuclear grade RTDs is 0.05°C. Better accuracies are difficult to achieve or maintain for any RTD that is built for industrial applications. The drift of nuclear grade RTDs was found to generally lie in a ± 0.2°C band. A drift band is used instead of drift rate because the drift of RTDs does not occur in a monotonic fashion to provide a unique value for calibration changes as a function of time.

The test results obtained in this project did not provide a reliable clue as to the useful life of nuclear plant RTDs. However, no reason was found to suggest that RTDs can not be used for their qualified life as specified by the manufacturers. This is provided that the RTDs have not exhibited any systematic or sudden drift or response time degradation as determined by periodic tests performed at least once during each operating cycle.

REFERENCES

1.   Hashemian, H. M., et.al., "Degradation of Nuclear Plant Temperature Sensors", U.S. Nuclear Regulatory Commission Report Number NUREG/CR-4928, June 1987.

2.   Kerlin, T. W., et.al., "In-Situ Response Time Testing of Platinum Resistance Thermometers", Electric Power Research Institute, Report Number NP-834, Volume 1, July 1978.

3.   Shepard, R. L.,et.al., "Remote Calibration of Resistance Temperature Devices", Electric Power Research Institute, Report Number NP-5537, February 1988.

4.   Riddle, John, et.al., "Platinum Resistance Thermometry", National Bureau of Standards, NBS Monograph 126, April 1973.

5.   Carr, K. R., "An Evaluation of Industrial Platinum Resistance Thermometers", Temperature Its Measurement and Control in Science and Industry, Volume 4, Part 2, Instrument Society of America, 1972.

6.   Mangum, B. W., "Stability of Small Industrial Platinum Resistance Thermometers", Journal of Research of the National Bureau of Standards, Volume 89, Number 4, July - August 1984.

7.   Hashemian, H. M., Petersen, K. M., "Calibration and Response Time Testing of Industrial RTDs", Proceedings of the 34th International Instrumentation Symposium, Instrument Society of America, May 1988.

8.   Hashemian, H. M., Petersen, K. M., "Effect of Aging on Performance of Nuclear Plant RTDs", Proceedings of the International Nuclear Power Plant Aging Symposium, Nuclear Regulatory Commission, NUREG/CP-0100, June 1988.

9.   "Search of the NPRDS Database for Failures of Nuclear Plant RTDs", Report Number NRC8904R0, Analysis and Measurement Services Corporation, December 1989.

10.   McGarry, J. T., Versluis, R. M., "RTD By-Pass Loop Elimination on Pressurized Water Reactors", International Conference on Operability of Nuclear Systems in Normal and Adverse Environments, Opera 89, Lyon, France, September 1989.

11.   Quinn, T. J., "News From the BIPM; Bureau International des poids et Measures", Metrologia 26, 69-74, (1989).

APPENDIX A

DESTRUCTIVE TESTING OF RTDs

APPENDIX A

DESTRUCTIVE TESTING OF RTDs

Qualitative tests were performed on 30 nuclear and 12 commercial grade RTDs under severe conditions to study tolerance to water intrusion, high temperature, mechanical shock, and thermal shock. A listing of the nuclear grade RTDs tested is given in Table A.1 along with the aging tests in the sequence performed. The same information for commercial grade RTDs is given in Table A.2. Any RTD whose element opened in these tests or exhibited more than 5°C shift in calibration was identified as a failed RTD.

The tests for tolerance to water intrusion involved submerging the RTDs in water at 50°C for 24 hours. This test resulted in one failure of a nuclear grade RTD, two failures of commercial grade RTDs, and small calibration changes in the remaining RTDs.

The high temperature tests were performed in a furnace at 650°C for eight hours. One nuclear and two commercial grade RTDs failed and the remaining RTDs suffered small calibration changes. The high temperature tests were followed by testing to simulate mishandling. Labeled as mechanical shock, the tests involved striking the RTDs against a wooden block about 50 times. This resulted in three failures in the nuclear grade RTDs and one failure in the commercial grade RTDs.

The thermal shock tests involved immersion of the RTDs into room temperature water after they had been inside a furnace for one hour at 650°C. This test produced the highest percentage of failures. The element resistance and the insulation resistance of the RTDs were monitored during the destructive tests to identify the failures as they occurred. There was little change in the insulation resistance except in some of the RTDs that failed. Figures A.1 through A.3 show the resistance changes for the RTDs during the mechanical shock tests. Three distinct behaviors were observed:

1. Resistance increased with the number of impacts (Figure A.1)
2. Resistance decreased with number of impacts (Figure A.2)
3. Resistance increased then decreased (Figure A.3)

The expected behavior is for the resistance to increase with the number of impacts. This is because mechanical shock is expected to cause work hardening in the platinum element. The unexpected behaviors seen in these tests could not be correlated with the changes in the insulation resistance or other diagnostic results for the RTDs.

A summary of the tests results are given in Table A.3. This is followed by Table A.4 with information about the RTDs that failed in the destructive tests. Note that open circuit is the dominant failure mode in these results.

TABLE A.1

Nuclear Grade RTDs Used in Destructive Tests

Tag	Water Intrusion	High Temperature	Mechanical Shock	Thermal Shock
3		◊		◊ ♦
4A			◊	
4C			◊	
5A			◊	
5C			◊	
7			◊	
9A	◊	◊	◊	
9C	◊	◊	◊	
11A	◊	◊	◊	◊ ♦
11C	◊	◊	◊	◊
12A	◊	◊		◊
12C	◊	◊		◊ ♦
13A	◊	◊		◊ ♦
13C	◊	◊		◊ ♦
14		◊		◊
15A			◊ ♦	
15C			◊ ♦	
16A			◊	
16C			◊	
17A	◊	◊		◊
17C	◊	◊		◊
18	◊	◊ ♦		
19	◊ ♦		◊	
20	◊	◊	◊ ♦	◊
21			◊	
22A			◊	
22C			◊	
23			◊	
24		◊		◊

◊ *Denotes RTDs tested*
♦ *Denotes RTDs failed*

TABLE A.2

Commercial Grade RTDs Used in Destructive Tests

Tag	Water Intrusion	High Temperature	Mechanical Shock	Thermal Shock
26		◊	◊	◊
27	◊	◊	◊	
35			◊	
36	◊	◊	◊	◊
38			◊	
42	◊ ♦		◊	
43	◊			
44	◊	◊ ♦	◊	
51	◊	◊ ♦	◊	◊
52			◊ ♦	
53	◊ ♦	◊	◊	
54	◊	◊		◊ ♦

◊ *Denotes RTDs tested*
♦ *Denotes RTDs failed*

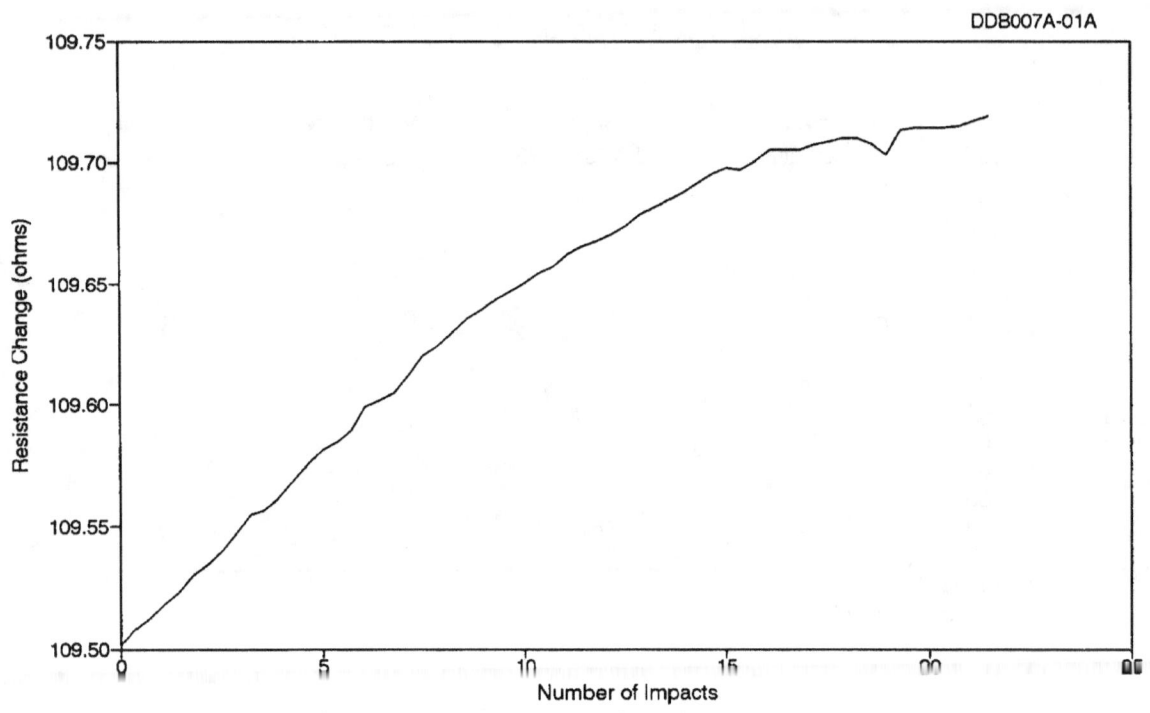

Figure A-1.   RTD Resistance Increasing During Tolerance
Tests for Mechanical Shock.

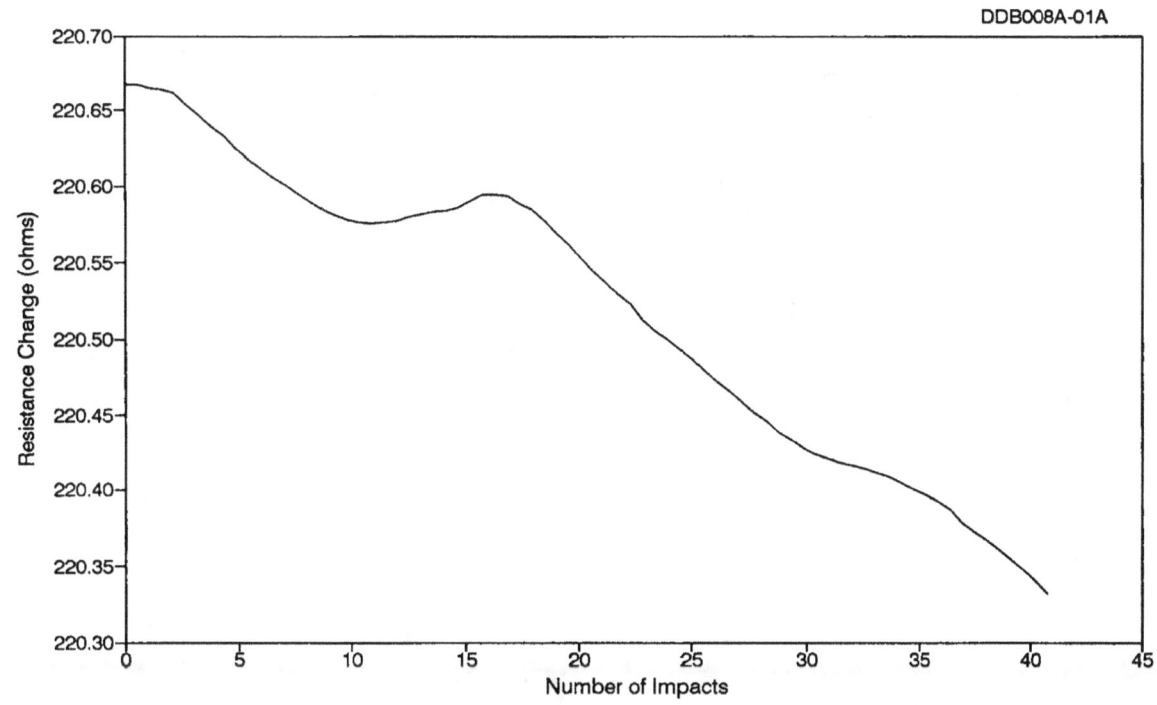

Figure A-2.   RTD Resistance Decreasing During Tolerance
Tests for Mechanical Shock.

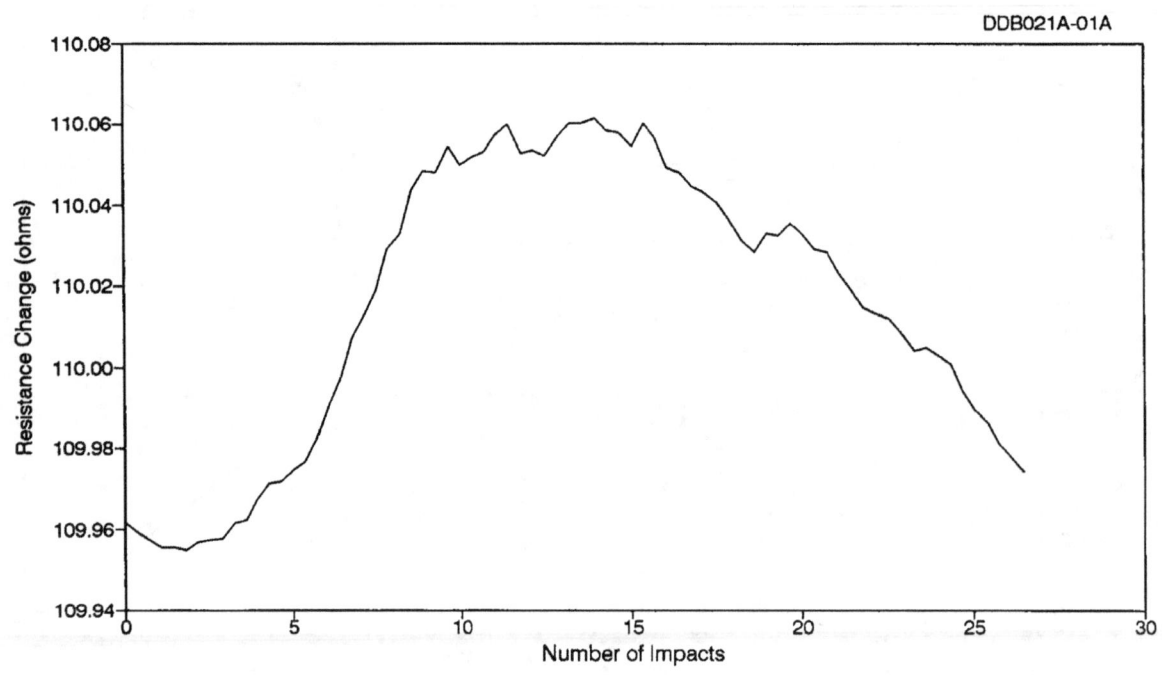

Figure A-3.   RTD Resistance Increasing and Decreasing During
                  Tolerance Tests for Mechanical Shock.

TABLE A.3

Summary of Destructive Test Results

	Water Intrusion	High Temperature	Mechanical Shock	Thermal Shock
Number Tested	21	22	29	16
Number Failed	3	3	4	6
Percent Failure	14%	14%	14%	38%

Results Summary

Total Number of RTDs Tested	=	41
Total Number of RTDs Failed	=	16
Percent of Total Failures	=	39%
Percent of Failures of Nuclear Grade RTDs	=	34%
Percent of Failures of Commercial Grade RTDs	=	50%

TABLE A.4

RTD Failure Summary

Tag	Aging Process	Reason for Failure	Remarks
19	Water Intrusion	Low IR	Recovered after a few weeks. IR back to 3GΩ
18	High Temperature	> 5°C Shift	_____
15A	Mechanical Shock	Open Circuit	Compensation leads were open.
15C	Mechanical Shock	Open Circuit	Compensation leads were open.
20	Mechanical Shock	Open Sensing Element	IR = 150 K Ω
3	Thermal Shock	> 5°C Shift	Large R_0 Shift. IR is reasonable.
11A	Thermal Shock	Open Circuit	_____
12C	Thermal Shock	Open Circuit	_____
13A	Thermal Shock	Open Circuit	_____
13C	Thermal Shock	Open Circuit	_____

IR = Insulation Resistance

APPENDIX B

SEARCH OF LER DATABASE

REVIEW OF LER DATABASE FOR RTD FAILURES
March 1990

This report describes a review of temperature instrumentation problems reported in Licensee Event Reports (LERs). The LERs reviewed in this study cover the period of 1970 to 1988. The data for the 1970 to 1983 period was obtained from a July 26, 1983 draft memorandum entitled "Survey of Temperature Related Licensee Event Reports" by R. M. Carroll of Oak Ridge National Laboratory. This report found a total of 412 LERs reporting failures of temperature measurement systems including RTDs and thermocouples. Of this, 71 reports were specific on RTDs. The LERs for the 1983 to 1989 period were identified using the Sequence Coding and Search System (SCSS) LER database operated by the Oak Ridge National Laboratory (ORNL) for the Nuclear Regulatory Commission (NRC).

About 950 reports were identified of which only 15 had specific reference to RTDs. LERs have been required of the nuclear industry since the early 1970s to report certain types of problems affecting nuclear power plants. From that time through 1983, the information reported in LERs remained relatively constant. Effective in 1984, new LER reporting requirements significantly reduced the number of LERs submitted. In general, LERs were no longer required for problems affecting only a single component. Such component failures were to be voluntarily reported to the Nuclear Plant Reliability Data System (NPRDS) operated by the Institute of Nuclear Power Operations (INPO). The consequence of the new LER reporting criteria as it affects temperature instrumentation was to reduce the number of LERs submitted solely to report single instrument failures.

Temperature instrumentation problems were reported in about 950 LERs. Of these, approximately 300 reported temperature switch problems, 300 reported temperature indicator problems, 200 reported problems with primary sensing elements, and about 150 reported problems with controllers, recorders, or transmitters. Only 15 of these reports specifically mentioned RTD failures. Together with R.M. Carroll's report, we found 92 specific RTD failures. These are categorized in the following table:

LER for 1970 - 1988 Period
Specifically Mentioning RTD Failures

Description	No. of Reported Failures	% of Total
Circuit Defect	32	35
Drift/Calibration	15	16
Moisture	15	16
Bad Connection/Wiring	15	16
Response Time	9	10
Other	6	7

Examples of LER abstracts are given in Table 1.

TABLE 1

Examples of RTD Problems Reported in LER Database

LER NUMBER	PLANT/UNIT	EVENT DATE	DESCRIPTION OF PROBLEM
247/81-026	INDIAN POINT 2	10/29/81	A DEFECT IN THE HOT LEG ROSEMOUNT ENGINEERING MODEL 176JA RESISTANCE TEMPERATURE DETECTOR CIRCUIT WAS THE CAUSE OF A REDUCTION IN DELTA AND AVERAGE TEMPERATURE IN LOOP 23.
247/87-020	INDIAN POINT 2	12/31/87	STATION PERSONNEL COULD NOT COMPLETE AN RTD HOOKUP TO A TERMINAL BLOCK DUE TO INSUFFICIENT LEAD LENGTH SUPPLIED WITH A NEW RTD. INVESTIGATION INTO EXTENDING THE RTD LEADS TO MAKE THE HOOKUP REVEALED THAT THE MANUFACTURER'S (RDF) VAPOR TIGHT REQUIREMENT FOR THESE LEADS WAS NOT MET. ON DECEMBER 31, 1987 THE RESULTS OF AN ENGINEERING REVIEW OF THE RDF WIDE RANGE REACTOR COOLANT SYSTEM (RCS) HOT LEG AND COLD LEG RTD'S INDICATED THAT NON-VAPOR TIGHT LEADS COULD HAVE COMPROMISED THEIR ENVIRONMENTAL QUALIFICATION.
272/88-002	SALEM 1	02/18/88	THE TAVG CONTROL ROOM INDICATION WAS IDENTIFIED TO REGISTER APPROXIMATELY 5F ABOVE THE WIDE RANGE REACTOR COOLANT SYSTEM (RCS) INDICATION. SUBSEQUENT INVESTIGATION REVEALED THE RESISTANCE COMPENSATION CIRCUIT FOR THE FIELD WIRING LENGTH (RTD TO THE LOW LEVEL AMP) WAS JUMPERED OUT NOT IN ACCORDANCE WITH DESIGN.
280/81-023	SURRY 1	07/08/81	DELTA-T PROTECTION CHANNEL T-1-422A WAS FOUND TO INDICATE ZERO. THE TC AND THE RTD'S WERE WIRED INCORRECTLY FOLLOWING COMPLETION OF A DESIGN CHANGE.
289/80-015	THREE MILE ISL 1	07/21/80	FOR APPROXIMATELY 6 HOURS THE ACTUAL RIVER WATER DELTA-T EXCEEDED THE -3F LIMIT, REACHING A MAXIMUM OF -5F. THIS WAS CAUSED BY INACCURATE COMPENSATION FOR THE RTD LEAD RESISTANCE DURING CALIBRATION.
295/80-022	ZION 1	05/07/80	THE OUTPUT SIGNAL FROM RC NARROW RANGE RTD 1TE-421A WAS INTERMITTENTLY SPIKING LOW. THE CAUSE WAS A BROKEN DRAIN WIRE AT THE WIRING CONNECTION TO THE RTD.
295/82-047	ZION 1	12/05/82	DELTA-T PEGGED LOW AND T-AVG PEGGED HIGH DUE TO THE SUDDEN FAILURE OF AN RTD DURING OPERATIONS. THE FAILED RTD, ROSEMOUNT 176KF, WAS REPLACED DURING NORMAL UNIT SHUTDOWN AND A VISUAL EXAMINATION PERFORMED TO IDENTIFY THE FAILURE MECHANISM. NO ANOMALIES WERE OBSERVED WHICH MIGHT INDICATE THE REASON FOR THE SUDDEN FAILURE OF THE RTD.
304/86-016	ZION 2	06/27/86	A LIGHTNING SURGE FOLLOWED A PATH FROM THE CONTAINMENT LINER TO GROUND VIA THE ELECTRICAL CABLE PENETRATIONS. THE CURRENT INDUCED IN THE CABLES WAS OF SUFFICIENT MAGNITUDE TO DAMAGE 5 HOT LEG TEMPERATURE RESISTANCE TEMPERATURE DETECTORS (RTD'S).
309/83-010	MAINE YANKEE	04/07/83	A NON-QUALIFIED RCS LOOP 2 RTD INPUT TO THE SUBCOOLING MARGIN MONITOR FAILED. CAUSE UNKNOWN.

313/88-001	ARKANSAS 1	12/17/88	A REACTOR COOLANT SYSTEM (RCS) HOT LEG RESISTANCE TEMPERATURE DETECTOR HAD BEEN CALIBRATED USING THE WRONG CALIBRATION DATA SINCE IT WAS INSTALLED IN MAY, 1982. THE CAUSE OF THIS EVENT WAS AN INADEQUATE PROCEDURE THAT DID NOT REQUIRE VERIFICATION OF THE COMPONENT SERIAL NUMBER UPON INSTALLATION.
315/80-002	COOK 1	01/25/80	ROSEMOUNT MODEL 176 KF RTD WAS PRODUCING ERRONEOUS TEMPERATURE READINGS. CAUSE NOT STATED.
315/81-007	COOK 1	03/18/81	SOSTMAN MODEL 119018 RTD PRODUCED ERRONEOUS TEMPERATURE READINGS. CAUSE NOT STATED.
317/80-067	CALVERT CLIFFS 1	12/21/80	TEMPERATURE ELEMENT 1-TE-122CB HAD FAILED. ITS PADDING REGISTER LEAD WAS OPEN. IT IS CONCLUDED THAT THE LEAD BROKE AS A RESULT OF HANDLING DURING INSTALLATION. ALSO, RTD 1-TE-112CB WAS FOUND WITH A HIGH RESISTANCE CONNECTION. BOTH TYPE 104ABH ELEMENTS WERE REPLACED WITH CALIBRATED SPARES.
317/81-009	CALVERT CLIFFS 1	02/05/81	TEMPERATURE DETECTOR Q-TE-112CD (ROSEMOUNT TYPE 104ABH) RESISTANCE HAD INCREASED BY NEARLY 1 OHM. THIS MAY BE DUE TO INCREASED CONNECTION RESISTANCE IN CONTAINMENT.
317/81-016	CALVERT CLIFFS 1	03/05/81	TCOLD INPUT TE-122CB WAS READING ERRATIC. TEMPERATURE DETECTOR 1-TE-122CB (ROSEMOUNT TYPE 104ABH) RESISTANCE WAS FOUND TO BE CHANGING. POSSIBLE CAUSES ARE THE ELEMENT OR ITS ELECTRICAL CONNECTIONS WITHIN THE CONTAINMENT.
317/82-041	CALVERT CLIFFS 1	07/13/82	TEMPERATURE DETECTOR 1-TE-112HA (ROSEMOUNT MODEL 104-1713-1) RESISTANCE WAS FOUND TO BE CHANGING. POSSIBLE CAUSES ARE THE ELEMENT OR ITS ELECTRICAL CONNECTIONS IN THE CONTAINMENT.
317/82-057	CALVERT CLIFFS 1	09/27/82	TEMPERATURE DETECTOR 1-TE-112HA (ROSEMOUNT TYPE 104ABH) RESISTANCE HAD INCREASED BY NEARLY 2 OHMS. PROBABLE CAUSE IS A LOOSE CONNECTION IN CONTAINMENT.
318/81-039	CALVERT CLIFFS 2	08/10/81	TEMPERATURE DETECTOR 2-TE-122CD (ROSEMOUNT TYPE 104ABH) RESISTANCE WAS FOUND TO BE CHANGING. POSSIBLE CAUSES ARE THE DETECTOR OR LOOSE ELECTRICAL CONNECTIONS WITHIN THE CONTAINMENT.
318/81-043	CALVERT CLIFFS 2	08/20/81	RESISTANCE TEMPERATURE DETECTOR (ROSEMOUNT TYPE 104ABH) WAS INTERMITTENTLY FLUCTUATING IN VALUE. PROBABLE CAUSE IS A VIBRATION-INDUCED OPEN ELEMENT.
327/80-055	SEQUOYAH 1	05/13/80	THE LOOP 2 HOT LEG RTD WAS FOUND TO BE READING ABOUT 3F IN ERROR. THE MALFUNCTION WAS CAUSED BY A HIGHER THAN NORMAL RESISTANCE BETWEEN THE RTD AND THE PROCESS INSTRUMENT RACK.
336/81-039	MILLSTONE PT 2	12/03/81	CHANNEL 'B' OF THE RPS BECAME ERRATIC DUE TO A FAILURE OF THE COLD LEG TEMPERATURE (TC) INPUTS. THE TC FAILURE RESULTED FROM MOISTURE FROM A STEAM LEAK ACCUMULATING ON THE RTD LEADS IN A TERMINAL BOX THAT HAD BEEN LEFT OPEN.
336/83-013	MILLSTONE PT 2	03/29/83	THE RESPONSE TIME FOR THE LOOP #1 HOT LEG, ON CHANNEL D, WAS FOUND TO BE OUT OF SPECIFICATION IN A NONCONSERVATIVE DIRECTION. THE CAUSE WAS FAILURE OF A ROSEMOUNT MODEL 104AFC-1 RTD.

336/84-006	MILLSTONE PT 2	02/13/84	A TOTAL OF 16 RTD'S WERE TESTED, OF WHICH 12 EXCEEDED THE TECH SPEC LIMIT OF 8 SECS OR LESS. SEVERAL FACTORS WERE DETERMINED TO CONTRIBUTE TO THE LARGE TIME CONSTANTS FOR THE RTD'S. THESE FACTORS INCLUDED THE TIME CONSTANT OF THE RTD ELEMENT ITSELF, THE INSTALLATION OF THE RTD IN THE THERMOWELL (INSUFFICIENT INSERTION DEPTH), AND A SENSOR-THERMOWELL MISMATCH.
338/81-074	NORTH ANNA 1	09/04/81	ERRATIC DELTA-T/TAVG READINGS BELIEVED TO BE CAUSED BY A CABLE TERMINATION PROBLEM AT THE T-COLD RTD WERE OBSERVED.
338/85-027	NORTH ANNA 1	12/24/85	CHANGE IN THE RESISTANCE CHARACTERISTICS OF THE LOOP B COLD LEG RESISTANCE TEMPERATURE DETECTOR (RTD) WERE OBSERVED.
339/80-007	NORTH ANNA 2	05/21/80	RTD FAILED. CAUSE NOT DETERMINED.
339/80-100	NORTH ANNA 2	12/11/80	DRIFT OF ROSEMOUNT RTD.
339/81-056	NORTH ANNA 2	07/02/81	ROSEMOUNT RTD FAILURE. CAUSE NOT KNOWN.
361/83-086	SAN ONOFRE 2	07/28/83	SUBCOOLED MARGIN MONITOR (SMM) "A" FAILED TO ZERO. INVESTIGATION REVEALED AN OPEN RESISTANCE TEMPERATURE DETECTOR (RTD) LEAD FOR THE LOOP 1, COLD LEG TEMPERATURE ELEMENT (2TE0911Y1) TO SMM "A".
362/83-092	SAN ONOFRE 3	10/15/83	FAULTY T-COLD RTD. CAUSE NOT STATED.
368/81-017	ARKANSAS 2	04/24/81	29 OUT OF 32 RTD'S DID NOT MEET THE REQUIRED RESPONSE TIME OF 6.0 SECONDS. INVESTIGATION HAS DETERMINED THAT THE COUPLANT USED IN RTD WELLS DRIED OUT DURING OPERATION AND THUS DID NOT PROVIDE ADEQUATE CONTACT OF RTD TO THERMOWELL WALL.
368/82-001	ARKANSAS 2	01/07/82	SIX (6) REACTOR COOLANT SYSTEM (RCS) RESISTANCE TEMPERATURE DETECTORS (RTD'S) DID NOT MEET THE REQUIRED RESPONSE TIME OF 6.0 SECONDS. THE COUPLANT USED IN THE RTD WELLS DRIED OUT DURING OPERATION AND DID NOT PROVIDE ADEQUATE CONTACT OR COUPLING BETWEEN THE RTD AND THERMOWELL WALL.
368/84-009	ARKANSAS 2	03/19/84	THE RESPONSE TIME OF ONE RTD SUPPLYING RPS CHANNEL "D" COLD LEG TEMPERATURE INDICATION WAS BEYOND THE TECH SPEC ALLOWABLE VALUE. DEGRADED RESPONSE TIME IS BELIEVED TO BE CAUSED BY INADEQUATE THERMAL COUPLING BETWEEN THE RTD SENSING ELEMENT AND RTD THERMOWELL. THE AFFECTED RTD'S ARE A MODEL 104AF, MANUFACTURED BY ROSEMOUNT, AND MODEL N9004, MANUFACTURED BY WEED.
382/86-005	WATERFORD 3	03/18/86	HOT LEG OUTLET TEMPERATURE RESISTANCE TEMPERATURE DETECTOR (RTD), FAILED (DUE TO CONNECTION EMBRITTLEMENT).
389/83-057	ST LUCIE 2	09/21/83	CHANNEL "C" B HOT LEG TEMPERATURE INSTRUMENTATION FAILED DUE TO INTERMITTENT GROUNDS ON ONE OF THE RDF CORP. MODEL 21251-SL2 RTD'S.
455/87-001	BYRON 2	01/15/87	A REACTOR TRIP OCCURRED DUE TO A REACTOR COOLANT RESISTANCE TEMPERATURE DETECTOR (RTD) FAILURE. THE FAILED RTD WAS BENCH TESTED AND DETERMINED TO BE OPERABLE. THE ACTUAL FAILURE WAS BELIEVED TO BE CAUSED BY A POOR SPLICE CONNECTION BETWEEN THE RTD AND THE FIELD CABLE.

APPENDIX C

SEARCH OF NPRDS DATABASE

**ANALYSIS AND
MEASUREMENT SERVICES
CORPORATION**

9111 CROSS PARK DRIVE NW / KNOXVILLE, TN 37923-4599 / (615) 691-1756

Report # <u>NRC8904R1</u>

<u>Final Report</u>

SEARCH OF THE NPRDS
DATABASE FOR FAILURES
OF NUCLEAR PLANT RTDS

Revision 1
March 1990

<u>Prepared for</u>

U.S. Nuclear Regulatory Commission
Office of Nuclear Regulatory Research
Washington, D.C. 20555

Contract No. <u>04-87-372</u>
FIN No. <u>D-2039</u>

NOTE

This is a condensed version of a confidential AMS report (No. NRC8904R0) submitted to the NRC in December 1989. The information presented herein is a compilation of the NPRDS data provided by the NRC on computer disks. All references to plant names and RTD manufacturers were deleted in presenting this information except for any mention of the manufacturer's name in the NPRDS failure narratives.

1.0 INTRODUCTION

This report presents a review of the Nuclear Plant Reliability Data System (NPRDS) database for failures of resistance temperature detectors (RTDs). The NPRDS database is operated by the Institute of Nuclear Power Operations (INPO) - an industry sponsored organization. Failure data for the RTDs comprises a small part of the NPRDS database. The NPRDS database contains detailed information describing failures of a broad range of components primarily for safety-related systems. The information on the component failures is submitted voluntarily to the NPRDS. Failure reports date back to the mid-1970's.

The purpose of this review was to identify any trend in degradation of nuclear plant RTDs and determine any dominate failure modes.

2.0 EVALUATION OF NPRDS DATA

NPRDS failure reports are supposed to be submitted when a component failure results in the failure of a reportable system to operate properly. The system's operability must be either lost or sufficiently degraded to inhibit proper function. Typically, instrumentation channels are provided in redundancy such that failures of individual components do not result in the loss of operability of entire systems. Therefore, single failures of instruments - whether from component mechanical defects or from calibration problems - are not reportable to NPRDS.

The NPRDS database contained information on 318 RTD failures as of May 1989. These failures were retrieved by searching for the component code for transmitter (IXMITR), engineering codes for a component type of sensor, and a principle of operation of resistance change. A breakdown of the number of failures by year is given in Table 1. As can be seen in the table, the number of failures voluntarily reported to NPRDS was quite small until 1984. The increase in failures during 1984 is probably the result of a concerted effort by INPO to improve the quantity and quality of data being reported - not just for RTDs, but for all components. At present, there is also discussion to broaden the scope of component failure data reportable to INPO to include certain secondary system failures: currently, only component failures in primary systems or systems essential to safety are reportable.

Although the number of RTD failures reported to NPRDS increased by almost an order of magnitude in 1984, the absolute number of failures reported is still relatively small - on average, substantially smaller than one failure per year for each of the approximately 100 operating plants. It is not clear from the data whether this small number is due to the reliability and dependability of RTDs or to infrequent reporting. Questions regarding the consistency of reporting hamper determinations about whether the increasing trends observed from 1984 through 1987 are due to a higher RTD failure rate or to changes in reporting.

Table 1 - NPRDS Reports/yr

Year	No. Reports
1974	3
1975	4
1976	6
1977	4
1978	2
1979	4
1980	2
1981	11
1982	5
1983	9
1984	38
1985	55
1986	59
1987	76
1988	37
1989	3

The table also shows fewer reports received in 1988 and 1989. Since the NRPDS search was performed in May 1989, a fewer number of 1989 reports is expected. Still, if the number of failures during 1989 is comparable to previous years, the small number of 1989 failures shown in the table, and also the fewer number for 1988, may be due to substantial lags between when a failure is discovered, documented, submitted, and ultimately added to the NPRDS database. These apparent time lags must be considered when searching the NPRDS database for current topics.

Tables 2 and 3 list individual and combined cause descriptions reported for the 318 RTD failures contained in the NPRDS database. (The cause descriptions sum to greater than 318 because multiple cause descriptions are frequently used on the NPRDS database records.) Based upon the NPRDS reportability criteria, these failures probably resulted in a loss of system function or a substantial degradation in system function. Simple calibration problems of single RTDs detected during periodic tests or inspections are not routinely reported to the NPRDS database.

The most frequently observed cause descriptions listed for the NRPDS failure records were: circuit defective, open circuit, normal/abnormal wear, calibration failure, short circuit, connection defective, and aging/cyclic fatigue. These seven account for about 73% of the cause descriptions. (A few samples of typical NPRDS failure reports for these cause codes are given in Tables 4 through 10 for the RTDs supplied by Conax, RdF, Rosemount, and Weed.) Except for the cause description out of calibration, the failure narratives generally reported that the problem was fixed by either repair or replacement activities. Even for those reported with the out of calibration category, the RTDs were replaced or repaired about half the time. A relatively small proportion of RTD failures were corrected by recalibrating or adjusting the RTD or associated instrumentation.

Table 2 - RTD Failure Cause Description Distribution

Cause Description	No. Reports
Abnormal Stress [1]	15
Aging/Cyclic Fatigue [1]	26
Burned/Burned Out [2]	5
Circuit Defective [2]	68
Connection Defective/Loose Parts [2]	31
Contacts Burned/Pitted/Corroded [1]	1
Corrosion [1]	11
Dirty [1]	7
Foreign/Incorrect Material [3]	2
Foreign/Wrong Part [3]	1
Incorrect Action [3]	12
Incorrect Procedure [3]	1
Insulation Breakdown [1]	11
Material Defect [2]	7
Mechanical Damage/Binding [2]	9
Normal/Abnormal Wear [1]	41
Open Circuit [2]	54
Previous Repair/Installation Status [3]	13
Out of Calibration [1]	36
Out of Mechanical Adjustment [1]	3
Particulate Contamination [1]	2
Setpoint Drift [1]	6
Short/Grounded [2]	34

Table 3 - Combined Categories

Cause Description	No. Reports
[1] Age-Related	159
[2] RTD or Circuit Defect	208
[3] Personnel Related	29

The remaining 23 cause descriptions accounted for about 27% of the data. As can be seen in Table 2 several of these descriptions were used infrequently.

Many of the cause descriptions listed in Table 2 may be indicative of problems associated with the effects of time. Time effects, or age effects, are particularly important when sensitive components are operated in environments of varying pressure, temperature, humidity, vibration, corrosion, and many other conditions which may serve to degrade RTD performance. The cause descriptions which are shown as reference number 1 in Table 2 are those potentially indicative of age-related conditions. They are summed in the age-related categories in Table 3. Similarly, summary categories for circuit defects and personnel errors are also shown. These combined categories are illustrated in Figure 1.

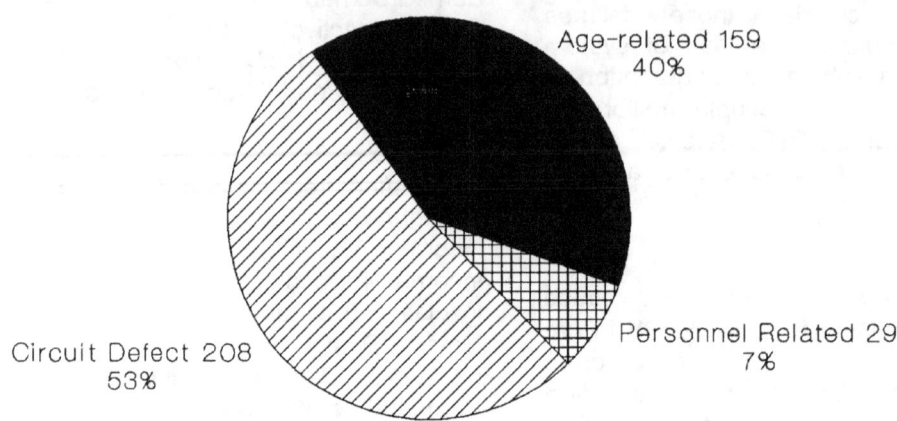

Figure 1 - Combined Cause Distribution

As can be seen, potential age-related causes are apparent in a substantial number of the NPRDS failure records.

3.0 CONCLUSIONS

A search of the NPRDS database for the period of 1974 to 1989 revealed 318 cases of failures of RTDs in nuclear power plants. Component failures are reportable to NPRDS whenever they lead to the failure or significant degradation of reportable systems. These systems are generally primary systems or those systems which fulfill an essential safety purpose. RTD failures in other systems or those failures which do not contribute to a reportable system failure are not presently reportable to NPRDS. Instrumentation systems for primary or essential safety systems are generally designed with redundancy such that single instrument failures would not degrade the entire system. For this reason, most instrument failures would not be reportable to NPRDS.

RTD failure reports were infrequently reported from the mid-1970s through 1983. The number of failures increased by about an order of magnitude in 1984. This is probably due to a concerted effort by INPO to increase both the quantity and quality of reporting to the NPRDS database - not just for RTD failures, but for all reportable component failures. Because of the changes in reporting to the NPRDS database, an increasing trend in the number of RTD failures from 1984 through 1987 cannot clearly be attributable to increasing problems with RTD reliability.

The most commonly reported cause descriptions for the RTD failures were: circuit defective, open circuit, normal/abnormal wear, out of calibration, short circuit, connection defective, and aging/cyclic failure. About 40% of the cause descriptions used in the NPRDS database for the RTD failures are indicative of age-related problems.

The dominant corrective actions taken for all of the cause descriptions (except for out of calibration) were repair or replacement. The corrective actions reported, in comparable numbers, for the out of calibration cause description were (1) calibration or adjustment and (2) repair or replacement.

During the period from 1984 through about May 1989, RTD failure reports were received from 52 plants. This is about half of the number of operating nuclear power plants. On average, about one report per operating plant was submitted about every two years.

TABLE 4

TYPICAL EXAMPLES OF FAILURES FROM DEFECTIVE CIRCUIT

Date of Event	Failure Narrative
10/17/80	DURING UNIT OPERATION, THE FIRE HAZARDS PANEL INDICATOR FOR THE 2C WIDE RANGE COLD LEG FAILED HIGH. THIS WAS DUE TO FAILURE OF THE 2C COLD LET RTD. THE RTD WAS FOUND TO BE DEFECTIVE. THE RTD WAS REPLACED AND TESTED.
7/16/81	DURING PLANT STARTUP THE RTD FOR CHANNEL B LOOP 3 COLD LEG WAS READING HIGH. FAULTY RTD. REPLACED RTD.
4/26/84	WITH UNIT AT 0% POWER, THE LOOP 1 HOT LEG TEMPERATURE WAS CAUSING HIGH LINEAR POWER DIFFERENTIAL AND LOW DEPARTURE FROM NUCLEAR BOILING (DNBR) TRIPS ON PLANT PROTECTIVE SYSTEM CHANNEL D. INVESTIGATION FOUND THAT TEMPERATURE ELEMENT WAS FAILING. SUSPECT RESISTANCE TEMPERATURE DETECTOR (RTD) WEAROUT. REPLACED RTD WITH IN KIND REPLACEMENT. PERFORMED CALIBRATION AND RESPONSE TIME TEST SATISFACTORILY. RETURNED TO SERVICE.
5/1/84	COMPUTER POINT A-1632, REACTOR COOLANT HOT LEG 'A' WATER TEMPERATURE, WAS READING 25 DEGREES LOWER THAN THE INDICATIONS FOR THE OTHER LOOPS. A DEFECTIVE RESISTANCE THERMAL DEVICE (RTD) WAS SENDING AN INCORRECT SIGNAL TO THE COMPUTER POINT. REPLACED THE RTD AND CHECKED THE OTHER COMPONENTS IN THE INSTRUMENT STRING SENDING A SIGNAL TO THE COMPUTER. THE STRING MET THE SPECIFICATIONS REQUIRED IN THE CONTROLLING PROCEDURE. COMPUTER POINT RETURNED TO SERVICE.
6/20/84	DURING THE PERFORMANCE OF THE NORMAL 18 MONTH RESISTANCE TEMPERATURE DETECTOR (RTD) TIME RESPONSE SURVEILLANCE, IT WAS DISCOVERED THAT THE 'A' CHANNEL COLD LEG RTD DID NOT MEET THE MINIMUM SURVEILLANCE REQUIREMENTS. THIS HAD NO IMMEDIATE EFFECT ON NORMAL PLANT OPERATIONS. THE CAUSE OF FAILURE IS UNKNOWN. DURING THE NEXT REFUELING OUTAGE A NEW RTD WAS CALIBRATED, INSTALLED AND RTD TE-1112CA WAS RETURNED TO SERVICE. PWO NO. 6654
6/24/85	REACTOR COOLANT LOOP "B" DELTA TEMPERATURE AVERAGE TEMPERATURE RESISTANCE TEMPERATURE DEVICE CAUSED A TRANSIENT IN REACTOR PROTECTION SYSTEM. UNIT WAS IN START UP. RESISTANCE TEMPERATURE DEVICE WAS READING HIGHER THAN OTHERS. RESISTANCE HIGHER THAN NORMAL. ROOT CAUSE OF FAILURE WAS UNKNOWN. REPLACED RESISTANCE TEMPERATURE DEVICE WITH A, LIKE FOR LIKE, SPARE ONE. (WR 124516)
07/30/85	WHILE THE PLANT WAS AT POWER, DURING A SURVEILLANCE TEST THE TEMPERATURE ELEMENT FOR THE REACTOR COOLANT SYSTEM HOT LEG WAS GIVING AN ERRATIC RESPONSE. THE RESISTANCE TEMPERATURE DETECTOR WAS DEGRADED. THE ROOT CAUSE IS UNKNOWN. REPLACED THE DETECTOR.
9/7/85	DURING POWER OPERATION AND MONTHLY REACTOR PROTECTION SYSTEM SURVEILLANCE, LOOP RESISTANCE TEMPERATURE DETECTOR (RTD) TE-132HD WAS FOUND TO PRODUCE INTERMITTENT SIGNALS. NOT YET DETERMINED. REPLACED THE RTD.
9/7/85	DURING MONTHLY REACTOR PROTECTION SYSTEM CALIBRATION AND POWER OPERATION, LOOP RESISTANCE TEMPERATURE DETECTOR (RTD) TE-132HB WAS FOUND TO PRODUCE INTERMITTENT SIGNALS. NOT YET DETERMINED. REPLACED THE RTD.
10/21/85	DURING REFUELING REACTOR PROTECTION LOGIN TESTS, PRIMARY COOLANT RESISTANCE TEMPERATURE DETECTOR 2-TE-402A WAS DISCOVERED WITH LOW SIGNAL WIRE TO GROUND RESISTANCE. IT IS UNKNOWN WHY THE RESISTANCES WOULD NOT FALL INTO MANUFACTURER SPECIFICATION. THE DETECTOR WAS REPLACED. THE INSTRUMENT CHANNEL WAS CALIBRATED, TESTED AND RESTORED TO SERVICE.

11/7/85	DURING POWER OPERATION AND MONTHLY REACTOR PROTECTION SYSTEM SURVEILLANCE, LOOP RESISTANCE TEMPERATURE DETECTOR (RTD) TE-132CD WAS FOUND TO PRODUCE INTERMITTENT SIGNALS. NOT YET KNOWN. REPLACED THE RTD.
11/9/85	DURING POWER OPERATION AND MONTHLY REACTOR PROTECTION SYSTEM SURVEILLANCE, LOOP RESISTANCE TEMPERATURE DETECTOR (RTD) TE-112CD WAS PRODUCING INTERMITTENT SIGNALS. NOT YET KNOWN. REPLACED THE RTD.
12/7/85	THE CONTROL ROOM REPORTED THAT ONE TEMPERATURE INPUT TO THE REACTOR PROTECTION SYSTEM FOR THE 1B1 COLD LEG TEMPERATURE WAS READING LOW. THE UNIT WAS AT FULL POWER AND THIS HAD NO IMMEDIATE EFFECT ON NORMAL PLANT OPERATIONS. IT WAS DETERMINED THAT TE-1122CA WAS INTERNALLY SHORTED. THE RESISTANCE TEMPERATURE DETECTOR (TE-1122CA) WAS CHANGED OUT WITH A NEW ONE FROM STORES AND TE-1122CA WAS RETURNED TO SERVICE. HAD TO WAIT UNTIL THE NEXT REFUELING OUTAGE. PWO NO. 8076
12/17/85	WHILE THE UNIT WAS STARTING UP, IT WAS NOTED THAT A TEMPERATURE ELEMENT FOR THE 'A' LOOP OF THE REACTOR COOLANT SYSTEM WAS NOT WORKING. THE ELEMENT WAS BYPASSED UNTIL AN OUTAGE. THE RESISTANCE TEMPERATURE DETECTOR WAS DEFECTIVE. IT WAS OPENING AT 125°F AND NOT REGISTERING. THE ROOT CAUSE HAS NOT BEEN DETERMINED. THE RTD WAS REPLACED. THE ELEMENT WAS THEN CALIBRATED AND VERIFIED OPERABLE.
1/7/86	DURING POWER OPERATION REACTOR PROTECTION SYSTEM (RPS) CHANNEL C DELTA POWER WAS SPIKING CAUSING PRETRIP ALARMS DUE TO FAILED RESISTANCE TEMPERATURE DETECTOR (RTD) TE-132CA. DR 0336-86 NORMAL END OF LIFE. REPLACED RTD.
1/8/86	WITH UNIT IN STARTUP OPERATIONS, MAINTENANCE NOTED THAT COLD LEG TEMPERATURE CHANNEL 'D' WAS NOT RESPONDING CORRECTLY. THIS IS A RESISTANCE TEMPERATURE DETECTOR (RTD) AND THIS DUAL ELEMENT SPARE WAS IN SERVICE. INVESTIGATION FOUND THAT DUAL RTD WAS FAILING. THE ONE IN USE FUNCTIONED FOR APPROXIMATELY 2 HOURS THEN FAILED. RTD HAD REACHED ITS END OF LIFE. REFERENCE NCR 3-1450. A FACILITY CHANGE NOTICE (FCN 455E, DCP3-6189. OE) WAS IMPLEMENTED. THE RTD WAS REPLACED WIT AN IN-KIND PART DURING A UNIT OUTAGE AND RETURNED TO SERVICE AFTER SATISFACTORY FUNCTIONAL TEST.
2/11/86	WITH UNIT 1 AT 100% POWER, AN ALARM IN THE CONTROL ROOM INDICATED THAT THE LOOP 'A' PROTECTION COLD LEG TEMPERATURE RESISTANCE TEMPERATURE DETECTOR (RTD) TE-1-412D HAD FAILED. THE CAUSE OF THE FAILURE HAS NOT BEEN DETERMINED. THE RTD WILL BE REPLACED DURING THE NEXT REFUELING OUTAGE.
10/08/86	PLANT STATUS AT REFUELING. TEMPERATURE READING FOR REACTOR COOLANT LOOP 2 HOT LEG WAS 51° ABOVE EXPECTED VALVE. DISCREPANCY NOTED DURING ROUTINE OBSERVATION BY CONTROL ROOM OPERATORS. NO EFFECT ON SYSTEM/PLANT OPERATIONS. DEFECTIVE RESISTANCE THERMAL DETECTOR WAS DISCOVERED. CAUSE OF FAILURE COULD NOT BE DETERMINED. REPLACED RESISTANCE THERMAL DETECTOR LIKE FOR LIKE. PERFORMED FUNCTIONAL TEST AND RETURNED LOOP TO SERVICE.
10/13/86	WITH THE UNIT AT 0% POWER, OPERATIONS NOTED THE REACTOR COOLANT LOOP 'A' COLD LEG TEMPERATURE WAS ACTING ERRATICALLY. THE RECORDER HAD SPURIOUS REVERSE DIFFERENTIAL TEMPERATURE, ALARM, AND INDICATION DIPPED AS LOW AS 200 DEGREES WITH THE ACTUAL TEMPERATURE AT 535 DEGREES F. INVESTIGATION DETERMINED EITHER THE TEMPERATURE ELEMENT OR THE ASSOCIATED CABLING HAD FAILED. DURING THE MIDCYCLE OUTAGE, THE CAUSE WILL BE INVESTIGATED FURTHER. SUSPECT WEAROUT. REFERENCE: NCR1-P-5993. THE TEMPERATURE ELEMENT WAS DISCONNECTED AND THE NEXT AVAILABLE ELEMENT WAS INSTALLED IN ITS PLACE. CALIBRATED AND RETURNED TO SERVICE.
11/29/86	DURING FULL POWER OPERATION, REACTOR COOLANT SYSTEM LOOP 'A' HOT LEG TEMPERATURE INDICATOR WAS DISCOVERED IN A FAILED HIGH CONDITION, WHILE OTHER INSTRUMENTS WERE OBSERVED TO BE NORMAL. THE INSTRUMENT CHANNEL DETECTOR HAD FAILED. THE DETECTOR WAS REPLACED. THE INSTRUMENT CHANNEL WAS CALIBRATED, TESTED AND RESTORED TO SERVICE.

12/18/86	REACTOR COOLANT HOT LEG LOOP "D" RESISTANCE TEMPERATURE DEVICE (RTD) WAS FAILING LOW IN THE CONTROL ROOM INDICATION. UNIT WAS IN POWER OPERATION. RESISTANCE TEMPERATURE DEVICE WAS GIVING ERRONEOUS OUTPUT, ROOT CAUSE UNKNOWN. CHANGED WIRING TO TAKE SIGNAL FROM SPARE RESISTANCE TEMPERATURE DEVICE, REPLACE RTD LIKE FOR LIKE AT A LATER DATE. (WR 129522, WR 67024)
12/19/86	REACTOR COOLANT LOOP 'D' HOT LEG RESISTANCE TEMPERATURE DEVICE (RTD) FAILED LOW GIVING LOW CONTROL ROOM INDICATION. UNIT AT FULL POWER. RESISTANCE TEMPERATURE DEVICE WAS DEFECTIVE AND NEEDED REPLACING, ROOT CAUSE UNKNOWN. (FURTHER INVESTIGATION PREVENTED BY ALARA) REPLACED DEFECTIVE RESISTANCE TEMPERATURE DEVICE. TIME RESPONSE TESTED NEW RTD. (WR 67024)
2/14/87	WITH THE UNIT IN HOT STANDBY, CONTROL ROOM OPERATIONS NOTED ERRATIC OPERATION OF THE WIDE RANGE "D" REACTOR COOLANT LOOP RESISTANCE TEMPERATURE DEVICE. LIMITING CONDITIONS OF OPERATION WERE ENTERED IN ACCORDANCE WITH TECH SPECS. TROUBLE SHOOTING WAS PERFORMED. THE CAUSE WAS UNKNOWN BUT ATTRIBUTED TO A DEFECTIVE CIRCUIT. THE DEFECTIVE RESISTANCE TEMPERATURE DEVICE WAS REPLACED AND THE INSTRUMENT LOOP WAS RECALIBRATED.
3/12/87	WITH UNIT AT 100% POWER, THE REACTOR COOLANT LOOP 1 HOT LEG TEMPERATURE INDICATION STARTED TO DRIFT. INVESTIGATION CONCLUDED THAT THE TEMPERATURE ELEMENT WAS DRIFTING AND AT ITS END OF LIFE. REFERENCE: NCR2-2001. REPLACED THE RESISTANCE TEMPERATURE DETECTOR WITH AN IN-KIND ONE. CALIBRATED AND VERIFIED PROPER OPERATION AND RETURNED TO SERVICE SATISFACTORILY.
5/30/87	2A REACTOR COOLANT SYSTEM COLD LEG WIDE RANGE TEMPERATURE INDICATOR 21F-413B METER AND CHART RECORDER SPIKED CAUSING CONTROL CIRCUITS TO ACTUATE, DURING NORMAL POWER OPERATION. THIS WAS DUE TO IMPROPER OPERATION OF THE ASSOCIATED RESISTANCE TEMPERATURE DETECTOR. RTD WAS DEFECTIVE. RTD WAS REPLACED.
8/7/87	WITH THE UNIT AT 92% POWER, OPERATIONS OBSERVED THE REACTOR COOLANT COLD LEG TEMPERATURE LOOP 'C' TO BE INDICATING 10 DEGREES LESS THAN NORMAL. READINGS WERE 510 TO 520 DEGREES INSTEAD OF THE NORMAL 520 TO 530 DEGREES. INVESTIGATION CONCLUDED THAT THE TEMPERATURE ELEMENT WAS FAILING AND THE RECORDER WAS FAILING. NORMAL WEAROUT/AGING OF THE TEMPERATURE ELEMENT. REFERENCE: NCR1-P-6209. REPLACED THE TEMPERATURE ELEMENT WITH A NEW ONE. TESTED AND RETURNED TO SERVICE. THE RECORDER WILL BE REPLACED AT A LATER DATE.
11/13/87	DURING SURVEILLANCE TEST, NO TEMPERATURE WAS INDICATED FOR THE REACTOR COOLANT LOOP '1B' COLD LEG TEMPERATURE CHANNEL 'C'. PLANT WAS AT FULL POWER. IT WAS SUSPECTED THAT THE TEMPERATURE ELEMENT FAILED DUE TO THE EXPOSURE TO HIGH TEMPERATURES OF NORMAL OPERATION. REPLACED THE TEMPERATURE ELEMENT AND CALIBRATED TRANSMITTER.
11/13/87	DURING REACTOR STARTUP OPERATIONS, REACTOR COOLANT BYPASS MANIFOLD RESISTANCE TEMPERATURE DETECTOR FAILED. THE CAUSE OF FAILURE IS UNKNOWN. THIS DETECTOR ALONG WITH ALL OTHER LOOP DETECTORS WERE TEMPORARILY REMOVED TO FACILITATE MODIFICATIONS TO THE BYPASS MANIFOLD PIPING DURING THE REFUELING OUTAGE. THE DETECTOR WAS REPLACED. THE NEW DETECTOR WAS CALIBRATED AND THE INSTRUMENT CHANNEL WAS RESTORED TO SERVICE.
7/9/88	DURING A REFUELING OUTAGE THE RPS (REACTOR PROTECTION SYSTEM) RTD (RESISTANCE TEMPERATURE DETECTOR) TIME RESPONSE PROCEDURE WAS BEING PERFORMED AND IT WAS DISCOVERED THAT TEMPERATURE ELEMENT TE-1112HC FOR REACTOR COOLANT LOOP NO. 1 HOT LEG DID NOT PASS ITS TIME RESPONSE TEST. NO EFFECT ON PLANT. THE CAUSE OF FAILURE IS UNKNOWN. A NEW RTD WAS INSTALLED, TESTED AND TE-1112HC WAS RETURNED TO NORMAL. PWO NO. 8022 PWO NO. 7732
7/9/88	DURING A REFUELING OUTAGE THE RPS (REACTOR PROTECTION SYSTEM) RTD (RESISTANCE TEMPERATURE DETECTOR) TIME RESPONSE PROCEDURE FOR CHANNEL "C" WAS BEING PERFORMED AND IT WAS DISCOVERED THAT TEMPERATURE ELEMENT TE-1122CC FOR REACTOR COOLANT LOOP 1B COLD LEG DID NOT PASS ITS TIME

RESPONSE TEST. NO EFFECT ON PLANT. THE CAUSE OF FAILURE IS UNKNOWN. A NEW RTD WAS INSTALLED AND TE-1122CC WAS RETURNED TO NORMAL. PWO NO. 8022 PWO NO. 7732

8/14/88

AT 100% POWER, FAILURE OF A REACTOR COOLANT SYSTEM WIDE RANGE TEMPERATURE DETECTOR CAUSED AN INTERMITTENT PRESSURE RELIEF VALVE LOW SETPOINT CUROUT ALARM. THE CAUSE OF THE TEMPERATURE DETECTOR FAILURE WAS UNKNOWN. THE TEMPERATURE DETECTOR WAS REPLACED AND A SATISFACTORY CALIBRATION WAS PERFORMED.

TABLE 5

TYPICAL EXAMPLES OF FAILURES DUE TO OPEN CIRCUIT

Date of Event	Failure Narrative
1/1/77	DURING OPERATION, TAVG CHANNEL TRIPPED LOW. RCS RTD LOOP RTD OPENED. IN SERVICE SPARE RTD WIRED TO CHANNEL, CHANNEL RECALIBRATED.
1/6/84	WITH UNIT AT POWER, OPERATIONS PERSONNEL RECEIVED AN ALARM INDICATING THAT THE REACTOR COOLANT SYSTEM LOOP "B" NARROW RANGE TEMPERATURE LOOP HAD FAILED DOWNSCALE. THE I&C DEPARTMENT DETERMINED THAT THE RESISTANCE TEMPERATURE DETECTOR (RTD) HAD FAILED OPEN. THE CAUSE OF FAILURE WAS UNKNOWN BUT COULD HAVE BEEN THE RESULT OF AGING CYCLIC FATIGUE. THE RTD WAS REPLACED. (84-278)
2/12/84	CHANNEL FAILED HIGH. DURING OPERATIONAL SURVEILLANCE OF CPC TRAIN "A" NOTED THAT INDICATOR 2TE91781 CHANNEL HAD FAILED HIGH. CAUSE: PROBABLY THERMAL CYCLING RESULTED IN INTERMITTENT OPEN AT RTD. FOUND INPUTS AT 2L121 TERMINALS 6, 7 & 8 HAD 5. 35 OHMS ON COMPUTER LEAD AND 00 AT ELEMENT. RTD OPENED UP WHEN HEATED ABOVE 400 DEGREES F. NCR 2-672, FCN S-95E, S96E. REPLACED SINGLE ELEMENT WITH DUAL ELEMENT RTD (MODEL NO. 104-AJA-1, ROSEMOUNT, RS03-P-1505-83).
9/27/84	WITH THE UNIT IN POWER OPERATION, OPERATIONS SHIFT PERSONNEL DETECTED A FAILURE ON "C" REACTOR COOLANT LOOP TEMPERATURE INSTRUMENTATION. INVESTIGATION REVEALED THAT THE "C" CONTROL LOOP TC COLD LEG RESISTANCE TEMPERATURE DETECTOR (RTD) HAD FAILED. THIS FAILURE DID NOT AFFECT THE SYSTEM OPERATION SINCE THE SPARE RTD WAS BROUGHT INTO SERVICE. THE CAUSE OF THE FAILURE WAS AN OPEN CIRCUIT ON THE "C" LOOP TC COLD LEG RTD WHICH RESULTED FROM NORMAL WEAR AND AGING. THE FAILURE WAS CORRECTED BY INSTALLING AND TESTING A NEW RTD IN ACCORDANCE WITH THE REPLACEMENT OF A NARROW RANGE REACTOR COOLANT RTD PROCEDURE AND THE INSTRUMENT CALIBRATION PROCEDURE FOR DELTA T/TAVG CALCULATION. THE SPARE RTD WAS RETURNED TO SPARE STATUS. 86-100
5/23/85	WITH THE UNIT AT 100% POWER, MAINTENANCE WAS WORKING ON OTHER SYSTEM RELATED EQUIPMENT AND FOUND THE LOOP 1 HOT LEG RESISTANCE TEMPERATURE DETECTOR (RTD) READING FULL SCALE HIGH. TROUBLESHOOTING FOUND THE RTD WITH AN OPEN CIRCUIT. PROBABLY THE RESULT OF AN INSTALLATION ERROR AND/OR A MANUFACTURING DEFECT. MOVED THE WIRING TO THE SPARE RTD (THIS IS A DUAL ELEMENT RESISTANCE TEMPERATURE DETECTOR). CALIBRATED, VERIFIED PROPER INDICATIONS AND RETURNED TO SERVICE.
7/30/85	WHILE THE PLANT WAS AT POWER, DURING A SURVEILLANCE TEST THE TEMPERATURE ELEMENT FOR THE REACTOR COOLANT SYSTEM HOT LEG WAS GIVING AN ERRATIC RESPONSE. THE RESISTANCE TEMPERATURE DETECTOR WAS DEGRADED. THE ROOT CAUSE IS UNKNOWN. REPLACED THE DETECTOR.
8/15/85	WITH THE UNIT IN COLD SHUTDOWN, OPERATIONS PERSONNEL ENCOUNTERED A FAILURE ON "B" LOOP INSTRUMENTATION. INVESTIGATION REVEALED THAT "B" LOOP COLD LEG RESISTANCE TEMPERATURE DETECTOR (RTD) HAD FAILED. THIS FAILURE DID NOT SIGNIFICANTLY EFFECT THE SYSTEM SINCE THE CHANNEL WAS DEFEATED. THE FAILURE WAS DUE TO AN OPEN CIRCUIT ON THE "B" LOOP COLD LEG RTD, CAUSE UNKNOWN. THE FAILURE WAS CORRECTED BY INSTALLING AND TESTING A NEW RTD IN ACCORDANCE WITH THE PROCEDURE FOR REPLACEMENT OF A NARROW RANGE REACTOR COOLANT RTD, AND THE INSTRUMENT CALIBRATION PROCEDURE FOR T421 DELTA T/TAVG CALCULATION. 86-098
8/27/85	DURING CALIBRATION OF LOOP B - PROTECTION INSTRUMENTATION CHANNEL 2, NO OUTPUT FROM LOOP LOW LEVEL AMPLIFIER COULD BE OBTAINED. REACTOR AT SHUTDOWN CONDITIONS. FAILED LOOP RESISTANCE TEMPERATURE DETECTOR (RTD)

TE-4-422D. DETECTOR CABLING WAS FOUND CRIMPED AND SQUASHED. NEW CABLING INSTALLED. RTD REPLACED. *7720-64*2 *7450-64*2

10/25/85

AT 100 PERCENT POWER ON OCTOBER 25, 1985, IT WAS OBSERVED THAT REACTOR COOLANT LOOP 1 DELTA T CONTROL INDICATOR 2RCTI2411B AND REACTOR COOLANT LOOP 1 T AVERAGE CONTROL INDICATOR 2RCTI2411A EXHIBITED NUMEROUS ERRATIC RESPONSES OF FULL SCALE, SHORT DURATION INDICATIONS. THE CAUSE OF THE FAILURE WAS FOUND TO BE AN OPEN REACTOR COOLANT LOOP 1 DELTA T/T AVERAGE CONTROL HOT LEG RESISTANCE TEMPERATURE DETECTOR 2RCTE2411B. DISCONNECTED THE OPEN TEMPERATURE DETECTOR 2RCTE2411B. JUMPERED REACTOR COOLANT LOOP 1 HOT LEG SPARE RESISTANCE TEMPERATURE DETECTOR 2RCTE2411D TO REPLACE 2RCTE2411B'S FUNCTION. THE DEFECTIVE DETECTOR WILL BE CHANGED DURING A FUTURE OUTAGE. RECALIBRATED AND RETURNED TO SERVICE SATISFACTORILY. WORK ORDER: 5900031627. DEVIATION REPORTS: 85-1399 AND 85-1412.

11/16/85

DURING POWER OPERATIONS, CONTROL ROOM PERSONNEL FOUND THE REACTOR WATER CLEANUP (RWCU) ROOM DIFFERENTIAL TEMPERATURE ALARM CARD READING DOWNSCALE. THIS DID NOT CAUSE ANY CHANGES IN OPERATING PARAMETERS. REF MWO 28504377. THE CAUSE OF THE FAILURE WAS FOUND TO BE A LOOSE CONNECTION ON THE RESISTANCE TEMPERATURE DETECTOR (RTD) THAT IS SENSED BY THE RWCU ROOM OUTLET TEMPERATURE ALARM CARD (G31-N661E). THE REASON FOR THE LOOSE CONNECTION WAS UNDETERMINED. THE CONNECTION WAS TIGHTENED, THE RTD WAS TESTED BY THE APPLICATION OF HEAT AND WAS RETURNED TO SERVICE.

2/21/86

WITH UNIT 1 AT 100% POWER, AN ALARM IN THE CONTROL ROOM ALERTED OPERATIONS. 'B' LOOP PROTECTION RESISTANCE TEMPERATURE DETECTOR (RTD), TE-1-422B, WAS FAILING WITH TAVE LOW AND DELTA T LOW. THE CAUSE OF THE FAILURE WAS DUE TO AN OPEN LEAD. THE RTD WAS TEMPORARILY JUMPERED.

2/25/86

REACTOR COOLANT LOOP "C" HOT LEG WIDE RANGE RESISTANCE TEMPERATURE DEVICE (RTD) WAS GIVING ERRONEOUS INDICATION IN THE CONTROL ROOM. THE RTD HAD FAILED HIGH. UNIT WAS ON LINE IN POWER OPERATION AT THE TIME. THE RESISTANCE TEMPERATURE DEVICE HAD A OPEN CIRCUIT IN IT AS MEASURED BY TECHNICIANS. ROOT CAUSE OF FAILURE WAS UNKNOWN. REPLACED RESISTANCE TEMPERATURE DEVICE LIKE FOR LIKE DURING OUTAGE. (W/R 122272, W/R 053791, W/R 053790, W/R 128169)

3/30/86

WITH THE UNIT IN HOT SHUTDOWN, SURVEILLANCE TESTING REVEALED AN OPEN CIRCUIT ON THE LOOP "C" COLD LEG RESISTANCE TEMPERATURE DETECTOR (RTD). THIS FAILURE HAD NO SIGNIFICANT EFFECT ON SYSTEM OPERATIONS BECAUSE THE SPARE RTD WAS UTILIZED AS A REPLACEMENT. THE CAUSE OF THE FAILURE IS UNKNOWN BUT THE OPEN CIRCUIT IS BELIEVED TO BE THE RESULT OF COMPONENT AGING. THE FAILURE WAS TEMPORARILY CORRECTED BY UTILIZING THE SPARE RTD VIA JUMPER UNTIL THE NEXT OUTAGE WHEN THE FAILED RTD WILL BE REPLACED. REPLACEMENT WORK TO BE PERFORMED USING WORK ORDER 041846. 86-246

5/1/86

WITH THE UNIT IN POWER OPERATIONS, SHIFT PERSONNEL DETECTED A FAILURE ON TH "C" LOOP DELTA T/TAVG PROTECTION INSTRUMENTATION. INVESTIGATION REVEALED A FAILED RESISTANCE TEMPERATURE DETECTOR (RTD). THIS FAILURE RESULTED IN THE LOSS OF A SYSTEM FUNCTIONAL PATH. THE EXACT CAUSE OF THE FAILURE IS UNKNOWN. THE OPEN CIRCUIT IS BELIEVED TO BE THE RESULT OF COMPONENT AGING. THE FAILURE WAS CORRECTED BY REPLACING THE FAILED RTD, CALIBRATING AND TESTING THE NEW RTD, AND RETURNING THE CIRCUIT TO SERVICE. 86-573

8/6/86

WITH THE UNIT IN HOT SHUTDOWN, SHIFT PERSONNEL DETECTED A FAILURE ON THE "A" LOOP DELTA T/TAVG PROTECTION INSTRUMENTATION. INVESTIGATION REVEALED A FAILED RESISTANCE TEMPERATURE DETECTOR (RTD). THIS FAILURE RESULTED IN THE LOSS OF A SYSTEM FUNCTIONAL PATH. THE EXACT CAUSE OF THE FAILURE IS UNKNOWN. THE OPEN CIRCUIT IS BELIEVED TO BE THE RESULT OF COMPONENT AGING. THE FAILURE WAS CORRECTED BY REPLACING THE FAILED RTD, CALIBRATING AND TESTING THE NEW RTD, AND RETURNING THE CIRCUIT TO SERVICE. 86-534

8/26/86

WITH THE UNIT AT 93% POWER, OPERATIONS PERSONNEL OBSERVED THE FAILURE OF THE "B" LOOP TEMPERATURE HOT - RESISTANCE TEMPERATURE DETECTOR (RTD) DUE TO ABNORMAL READINGS AND AUDIO VISUAL ALARM. THE FAILURE RESULTED IN THE

LOSS OF PROTECTION CHANNEL REDUNDANCY. THE CAUSE OF FAILURE WAS AN OPEN CIRCUIT IN A LEAD FROM THE RTD. A SPARE LEAD WAS CONNECTED AND THE INDICATION RETURNED TO NORMAL. 86-589

12/8/86   WHILE PLANT WAS OPERATING AT 100% POWER, THE I&C DEPARTMENT REPLACED THE RESISTANCE TEMPERATURE DETECTOR (RTD) AMPLIFIER CARD (NRA CARD) FOR THE 'C' LOOP TEMPERATURE HOT PROTECTION RTD. SOON AFTER THE CHANNEL AGAIN FAILED LOW. THIS HAD NO SIGNIFICANT PLANT OR SYSTEM AFFECT BECAUSE REDUNDANT CHANNELS WERE STILL IN SERVICE. THE EXACT CAUSE OF THE FAILURE WAS UNDETERMINED BUT ASSUMED TO BE THE RESULT OF A MANUFACTURING DEFECT COMMON TO THE MANUFACTURER (WEED) RTD'S. THE RTD CHANNEL WAS SWITCHED TO THE INSTALLED SPARE. 86-684

8/13/87   WITH UNIT 2 AT 99% POWER, 2B LOOP RESISTANCE TEMPERATURE DETECTOR (RTD) FAILED HIGH CAUSING CHANNEL NO. 2 OPDT, OTDT, ROD STOP, AND RUNBACK ANNUNCIATORS TO LOCK IN. (RING, WO: 055912, 87-086 0.4C2) OPEN LEAD ON RTD (TE-2-422B). PLACED CHANNEL IN TRIP. CHECKED SPARE RTD AND FOUND IT COMPLETELY DEFECTIVE. INSTALLED TEMPORARY MODIFICATION, ROLLING LEADS TERMINALS 1 AND 3. TESTED SATISFACTORY, RETURNED TO SERVICE. (NOTE: THIS MODEL HAS 4 LEADS AS INSTALLED, ONLY 3 REQUIRED.)

11/4/87   THE COMPUTER POINT A1692 FOR THE REACTOR PROTECTION SYSTEM CHANNEL "A" HOT LEG TEMPERATURE INDICATION WAS READING AN OPEN CONDITION. THE TECHNICIANS FOUND THAT RESISTANCE TEMPERATURE DETECTOR 1RC_RD0001A, USED TO MEASURE THE HOT LEG TEMPERATURE, HAD AN OPEN CIRCUIT. WE SUSPECT THAT AGING CONTRIBUTED TO THE COMPONENT FAILURE. THE TEMPERATURE DETECTOR WAS REPLACED AND ITS OPERATION WAS VERIFIED TO THE SPECIFICATIONS OF THE CONTROLLING PROCEDURE. THE COMPUTER READING RETURNED TO NORMAL.

6/15/87   WITH UNIT IN COLD SHUTDOWN, "C" LOOP COLD LEG NARROW RANGE (SPARE) RESISTANCE TEMPERATURE DETECTOR (RTD) WAS DETERMINED TO HAVE FAILED. THE FAILURE WAS INDICATED BY INFINITE RESISTANCE ACROSS TWO RESISTORS DURING PERFORMANCE OF THE RTD CROSS CALIBRATION PROCEDURE. EFFECT ON PLANT WAS MINIMAL BECAUSE ONLY THE SPARE RTD FAILED. FAILURE ANALYSIS HAS DETERMINED THAT THE COMPENSATION LEAD WAS OPEN RESULTING IN THE FAILURE, ROOT CAUSE UNKNOWN. RTD WAS REPLACED AND RECALIBRATE AT A LATER DATE. (87-401)

6/15/87   WITH UNIT SHUTDOWN, I&C PERSONNEL PERFORMING PREVENTIVE MAINTENANCE TESTING ON THE "B" COLD LEG NARROW RANGE SPARE RESISTANCE TEMPERATURE DETECTOR (RTD) DISCOVERED THAT IT WAS OUT OF SPECIFICATION. THERE WAS NO SIGNIFICANT EFFECT ON SYSTEM OPERATION BECAUSE THIS IS A SPARE RTD AND WAS NOT BEING USED. THE CAUSE OF FAILURE WAS INSULATION BREAKDOWN AND AN OPEN CIRCUIT BETWEEN COMPENSATING LEADS PER FAILURE ANALYSIS OF THE I&C DEPT. ROOT CAUSE OF FAILURE WAS UNKNOWN. THE RTD WAS REPLACED. (87-292)

4/27/88   WITH THE UNIT IN POWER OPERATION, THE RESISTANCE TEMPERATURE DETECTOR (RTD) FOR THE "A" REACTOR COOLANT LOOP HOT LEG TEMPERATURE PROTECTION CHANNEL FAILED. UNIT OPERATION WAS NOT AFFECTED AND ALL PROTECTION CHANNELS WERE TRIPPED TO ENSURE PLANT OPERATION CONTINUED IN A CONSERVATIVE PROTECTION MODE. AS RTD LEAD WIRE OPENED DUE TO POSSIBLE STRESS OF VIBRATION FROM FLOW OR TEMPERATURE CYCLING. THE SPARE RTD WIRE WAS CONNECTED AND THE CHANNEL WAS TESTED SATISFACTORILY AND RETURNED TO SERVICE. 88-136

TABLE 6

TYPICAL EXAMPLES OF FAILURES DUE TO NORMAL/ABNORMAL WEAR

Date of Event	Failure Narrative
1/6/84	WITH THE UNIT AT POWER OPERATION, PERSONNEL PERFORMING A PERIODIC TEST ON THE REACTOR COOLANT 'C' LOOP COLD LEG (SPARE) NARROW RESISTANCE TEMPERATURE DETECTOR (RTD) DISCOVERED THAT IT HAD SHORTED. THIS FAILURE HAD NO SIGNIFICANT EFFECT ON SYSTEM OPERATION BECAUSE ONLY THE SPARE RTD FAILED. THE EXACT CAUSE OF FAILURE WAS UNKNOWN, BUT IT WAS PROBABLY DUE TO NORMAL/ABNORMAL WEAR OF THE RTD CAUSING LOW INSULATION RESISTANCE. THE RTD WAS REPLACED, LEAK CHECKED, AND CALIBRATED. IT WAS THEN RETURNED TO NORMAL SERVICE. (84-343)
4/26/84	WITH UNIT AT 0% POWER, THE LOOP 1 HOT LEG TEMPERATURE WAS CAUSING HIGH LINEAR POWER DIFFERENTIAL AND LOW DEPARTURE FROM NUCLEAR BOILING (DNBR) TRIPS ON PLANT PROTECTIVE SYSTEM CHANNEL D. INVESTIGATION FOUND THAT TEMPERATURE ELEMENT WAS FAILING. SUSPECT RESISTANCE TEMPERATURE DETECTOR (RTD) WEAROUT. REPLACED RTD WITH IN KIND REPLACEMENT. PERFORMED CALIBRATION AND RESPONSE TIME TEST SATISFACTORILY. RETURNED TO SERVICE.
10/15/85	WITH UNIT 2 AT 100% POWER DURING NORMAL OPERATION, THE 'C' LOOP COLD LEG PROTECTION RESISTANCE TEMPERATURE DETECTOR (RTD), TE-2-432D FAILED. SPECIFIC CAUSE NOT DETERMINED. THE RTD WAS REPLACED WITH A SIMILAR RTD.
1/6/86	WITH THE UNIT AT 0% POWER, OPERATIONS NOTED THAT LOOP 2B COLD LEG THE TEMPERATURE CHANNEL 'D' TEMPERATURE ELEMENT FAILED CAUSING TEMPERATURE INDICATOR TO FAIL ALSO. INVESTIGATION FOUND A BAD TEMPERATURE ELEMENT. SUSPECT WEAROUT. REFERENCE: NCR3-1445. REPLACED THE TEMPERATURE ELEMENT WITH AN IN-KIND PART. TESTED AND PREFORMED LOOP VERIFICATION SATISFACTORILY, THEN RETURNED TO SERVICE.
1/7/86	DURING POWER OPERATION REACTOR PROTECTION SYSTEM (RPS) CHANNEL C DELTA POWER WAS SPIKING CAUSING PRETRIP ALARMS ·DUE TO FAILED RESISTANCE TEMPERATURE DETECTOR (RTD) TE-132CA. DR 0336-86 NORMAL END OF LIFE. REPLACED RTD.
1/8/86	WITH UNIT IN STARTUP OPERATIONS, MAINTENANCE NOTED THAT COLD LEG TEMPERATURE CHANNEL 'D' WAS NOT RESPONDING CORRECTLY. THIS IS A RESISTANCE TEMPERATURE DETECTOR (RTD) AND THIS DUAL ELEMENT SPARE WAS IN SERVICE. INVESTIGATION FOUND THAT DUAL RTD WAS FAILING. THE ONE IN USE FUNCTIONED FOR APPROXIMATELY 2 HOURS THEN FAILED. RTD HAD REACHED ITS END OF LIFE. REFERENCE NCR 3-1450. A FACILITY CHANGE NOTICE (FCN 455E, DCP3-6189. OE) WAS IMPLEMENTED. THE RTD WAS REPLACED WIT AN IN-KIND PART DURING A UNIT OUTAGE AND RETURNED TO SERVICE AFTER SATISFACTORY FUNCTIONAL TEST.
5/1/86	1RC_RD0001A IS A RESISTANCE THERMAL DEVICE (RTD) USED TO MEASURE TEMPERATURE IN THE 'A' LOOP OF THE REACTOR COOLANT SYSTEM. THE OUTPUT FROM THE DEVICE WAS LOWER THAN THE OTHER INSTRUMENTS IN THE CONTROL ROOM THAT PERFORM THE SAME FUNCTION. THE RTD OUTPUT WAS LOW. THIS WAS PROBABLY DUE TO NORMAL DETERIORATION OF THE ELEMENTS IN THE RTD. THE RTD WAS REPLACED WITH A NEW ONE. ALL THE INSTRUMENTS IN THIS STRING WERE CHECKED AND RECALIBRATED AS NECESSARY.
7/20/86	WITH UNIT 1 AT 54% POWER, CONTROL ROOM INDICATIONS SHOWED GREATER THAN 6% DIFFERENCE IN CHANNELS OF THE 'B' LOOP DELTA T PROTECTION TRANSMITTER (TE-1-422A). (86-063. 4C1 WO: 038745) FAILURE OF RESISTANCE TEMPERATURE DETECTOR (RTD) DUE TO AGE: REPLACED RTD.
10/13/86	WITH THE UNIT AT 0% POWER, OPERATIONS NOTED THE REACTOR COOLANT LOOP 'A' COLD LEG TEMPERATURE WAS ACTING ERRATICALLY. THE RECORDER HAD SPURIOUS REVERSE DIFFERENTIAL TEMPERATURE, ALARM, AND INDICATION DIPPED AS LOW AS 200

DEGREES WITH THE ACTUAL TEMPERATURE AT 535 DEGREES F. INVESTIGATION DETERMINED EITHER THE TEMPERATURE ELEMENT OR THE ASSOCIATED CABLING HAD FAILED. DURING THE MIDCYCLE OUTAGE, THE CAUSE WILL BE INVESTIGATED FURTHER. SUSPECT WEAROUT. REFERENCE: NCR1-P-5993. THE TEMPERATURE ELEMENT WAS DISCONNECTED AND THE NEXT AVAILABLE ELEMENT WAS INSTALLED IN ITS PLACE. CALIBRATED AND RETURNED TO SERVICE.

1/17/87   WITH UNIT AT COLD SHUTDOWN, PERSONNEL DISCOVERED DURING CALIBRATION TESTING THAT THE LOOP "A" HOT CONTROL RESISTANCE TEMPERATURE DETECTOR (TE-1-411B) HAD AN INSULATION BREAKDOWN AND WAS SENDING INCORRECT SIGNALS. (87-030. 4C1 WO: 048806) INSULATION BREAKDOWN DUE TO HEAT. INSTALLED NEW RESISTANCE TEMPERATURE DETECTOR.

3/12/87   WITH UNIT AT 100% POWER, THE REACTOR COOLANT LOOP 1 HOT LEG TEMPERATURE INDICATION STARTED TO DRIFT. INVESTIGATION CONCLUDED THAT THE TEMPERATURE ELEMENT WAS DRIFTING AND AT ITS END OF LIFE. REFERENCE: NCR2-2001. REPLACED THE RESISTANCE TEMPERATURE DETECTOR WITH AN IN-KIND ONE. CALIBRATED AND VERIFIED PROPER OPERATION AND RETURNED TO SERVICE SATISFACTORILY.

5/19/87   WITH UNIT 1 IN COLD SHUTDOWN DURING SURVEILLANCE TESTING, PERSONNEL FOUND THAT THE LOOP "A" SPARE RESISTANCE TEMPERATURE DETECTOR (RTD) (TE-1-411D) WAS READING OPEN INSTEAD OF CLOSED. (LASKOWSKI, WO: 053478, 87-081. 4C1) WEAR AND AGING ATTRIBUTED FROM SYSTEM STRESS. REPLACED RTD.

5/25/87   DURING NORMAL PLANT OPERATION, THE REACTOR COOLANT SYSTEM COLD LEG WIDE RANGE TEMPERATURE INDICATOR BEGAN FLUCTUATING. THE COMPONENT WAS REDUNDANT SO THERE WAS NO EFFECT ON THE UNIT. THE CAUSE OF THE EVENT WAS FAILURE OF TEMPERATURE ELEMENT 2TE-0413B DUE TO WEAR. THE TEMPERATURE ELEMENT WAS REPLACED.

6/4/87   DURING REFUELING SHUTDOWN AND ROUTINE TESTING, REACTOR PROTECTION SYSTEM LOOP 2 HOT LEG TEMPERATURE RESISTANCE TEMPERATURE DETECTOR (RTD) TE-122HD FAILED TO MEET THE REQUIRED RESPONSE TIME. DR 3600-87 NORMAL WEAROUT. REPLACED RTD.

7/13/87   UNIT 2 WAS OPERATING AT 100% POWER. CONTROL ROOM OPERATORS OBSERVED ERRATIC OUTPUT ON THE REACTOR COOLANT SYSTEM HOT LEG TEMPERATURE INDICATOR (TE-2-433). EXACT CAUSE WAS UNKNOWN, BUT SUSPECT NORMAL WEAR AND AGING OF TEMPERATURE TRANSMITTER (TE-2-433) REPLACED TRANSMITTER LIKE FOR LIKE.

6/21/88   REACTOR COOLANT LOOP "C" COLD LEG RESISTANCE TEMPERATURE DETECTOR (RTD) WAS READING LOW. FOUND BY OPERATOR WITH UNIT AT POWER. DEFECTIVE RTD READING 4 DEGREES LOW, ROOT CAUSE UNKNOWN. RTD TO BE REPLACED DURING UNIT REFUELING OUTAGE. (WR 135046)

TABLE 7

TYPICAL EXAMPLES OF CALIBRATION FAILURES

Date of Event	Failure Narrative
5/8/80	LOOP 4 TC INSTRUMENT READING LOW. INSTRUMENT DRIFT. CALIBRATED LOOP 4 TC INSTRUMENT.
7/5/85	REACTOR COOLANT LOOP "A" HOT LEG RESISTANCE THERMAL TRANSMITTING DEVICE APPEARED TO HAVE ONE LEAD GROUND. THE CIRCUIT WAS SEARCHED FOR GROUND. NO REASON COULD BE FOUND. THE TRANSMITTER TEMPORARILY WAS REPLACED. THE CIRCUIT THEN RESPONDED PROPERLY. THE ORIGINAL COMPONENT TESTED CORRECTLY AND WAS REINSTALLED. (WR 65487)
12/27/85	WITH PLANT STATUS AT 100% POWER, TEMPERATURE AVERAGE FOR REACTOR COOLANT LOOP 2 WAS DEVIATING FROM TEMPERATURE AVERAGE FOR LOOPS 1, 3, 4. DISCREPANCY NOTED BY CONTROL ROOM OPERATOR DURING ROUTINE OBSERVATION. NO EFFECT ON PLANT. RESISTANCE TEMPERATURE DETECTOR (RTD) WAS FOUND TO BE OUT OF CALIBRATION. RECALIBRATED RTD AND PERFORMED STRING TEST.
4/24/86	PLANT AT REDUCED POWER FOLLOWING ANNUAL REFUELING OUTAGE. REACTOR COOLANT HOT LEG RESISTANCE TEMPERATURE DETECTOR WAS VARYING BY 10 DEGF DIFFERENCE. OUT OF CALIBRATION. RECALIBRATED SATISFACTORILY PER PROCEDURE.
4/30/86	PLANT AT FULL POWER. REACTOR COOLANT COLD LEG RESISTANCE TEMPERATURE DETECTOR READ 19 DEGF BELOW WHAT IT SHOULD BE ACCORDING TO COMPUTER AND CONTROL ROOM RECORDER. OUT OF CALIBRATION. RECALIBRATED PER PLANT PROCEDURE.
10/2/86	DURING UNIT REFUELING WHILE PERFORMING ROUTINE CALIBRATION, THE 'A' STEAM TUNNEL HIGH TEMPERATURE TRANSMITTER WAS FOUND TO BE OUT OF SPECIFIED TOLERANCES. THERE WAS NO SIGNIFICANT EFFECT ON SYSTEM OPERATION. REF. MWO 2-86-5146. THE PROBLEM WAS ATTRIBUTED TO AN OUT-OF-CALIBRATION CONDITION OF THE TRANSMITTER AND COULD NOT BE CORRECTED BY ADJUSTMENT. THE TRANSMITTER FAILED DUE TO UNKNOWN INTERNAL CAUSES. THE DEFECTIVE TRANSMITTER WAS REPLACED WITH A NEW ONE. THE NEW TRANSMITTER WAS CALIBRATED PER PLANT PROCEDURE TO PROVE OPERABILITY.
11/26/86	UNIT AT 98% POWER, AN ALARM WAS RECEIVED ON COMPUTER POINT FROM THE REACTOR COOLANT LOOP "C" TEMPERATURE DETECTOR. RESISTANCE THERMAL DETECTOR WAS OUT OF CALIBRATION. RECALIBRATED RESISTANCE THERMAL DETECTOR AND RETURNED LOOP TO SERVICE.
11/26/86	PLANT STATUS AT 44% FULL POWER, REACTOR COOLANT LOOP "B" COLD LEG WIDE RANGE TEMPERATURE DETECTOR WAS OUT OF CALIBRATION. DISCREPANCY NOTED DURING SCHEDULED SURVEILLANCE TESTING. NO EFFECT ON PLANT. LOGIC CARD FOR RESISTANCE THERMAL DETECTOR WAS OUT OF CALIBRATION. RECALIBRATED LOGIC CARD. RETURNED WIDE RANGE TEMPERATURE DETECTOR LOOP TO SERVICE.
2/17/87	WITH UNIT AT POWER, OPERATIONS PERSONNEL RECEIVED A DEVIATION FROM DELTA T CALCULATED TEMPERATURE ALARM. INVESTIGATION BY THE I&C DEPARTMENT SHOWED THAT THE "C" HOT LEG NARROW RANGE RESISTANCE TEMPERATURE DETECTOR (RTD) WAS DRIFTING. THERE WAS NO SIGNIFICANT EFFECT ON SYSTEM OPERATION BECAUSE THE CHANNEL WAS SWITCHED TO A SPARE RTD. THE CAUSE OF FAILURE WAS UNKNOWN. THE RTD WAS REMOVED FROM SERVICE, THE LOOP WAS SWITCHED TO THE INSTALLED SPARE AND THE RTD WILL BE REPLACED AT THE NEXT OUTAGE. (87-369)
3/11/87	WITH UNIT AT POWER, I&C PERSONNEL PERFORMING SURVEILLANCE TESTING ON THE REACTOR COOLANT SYSTEM NARROW RANGE TEMPERATURE INSTRUMENTATION DISCOVERED THAT THE RESISTANCE TEMPERATURE DEFECTOR (RTD) WAS STARTING TO

DRIFT. THE CAUSE OF FAILURE WAS UNKNOWN. THE RTD WAS REPLACED AT THE FIRST SHUTDOWN AND THE LOOP WAS RECALIBRATED PER STATION PROCEDURES BEFORE BEING RETURNED TO SERVICE. (87-372)

5/25/87   DURING NORMAL PLANT OPERATION, THE REACTOR COOLANT SYSTEM COLD LEG WIDE RANGE TEMPERATURE INDICATOR BEGAN FLUCTUATING. THE COMPONENT WAS REDUNDANT SO THERE WAS NO EFFECT ON THE UNIT. THE CAUSE OF THE EVENT WAS FAILURE OF TEMPERATURE ELEMENT 2TE-0413B DUE TO WEAR. THE TEMPERATURE ELEMENT WAS REPLACED.

6/15/87   WITH UNIT SHUTDOWN, I&C PERSONNEL PERFORMING PREVENTIVE MAINTENANCE TESTING ON THE "B" COLD LEG NARROW RANGE SPARE RESISTANCE TEMPERATURE DETECTOR (RTD) DISCOVERED THAT IT WAS OUT OF SPECIFICATION. THERE WAS NO SIGNIFICANT EFFECT ON SYSTEM OPERATION BECAUSE THIS IS A SPARE RTD AND WAS NOT BEING USED. THE CAUSE OF FAILURE WAS INSULATION BREAKDOWN AND AN OPEN CIRCUIT BETWEEN COMPENSATING LEADS PER FAILURE ANALYSIS OF THE I&C DEPT. ROOT CAUSE OF FAILURE WAS UNKNOWN. THE RTD WAS REPLACED. (87-292)

6/17/87   WITH THE UNIT IN HOT STANDBY, OPERATING PERSONNEL NOTICED THE 2D WIDE RANGE RESISTANCE TEMPERATURE DETECTOR (RTD) TO BE FLUCTUATING 1 TO 2 DEGREES F. THIS PREVENTED AN INCREASE IN REACTOR POWER. RTD OUT OF CALIBRATION, CAUSE UNKNOWN. THE RESISTANCE TEMPERATURE DETECTOR WAS REPLACED AND A 1 POINT AMBIENT CHECK WAS PERFORMED.

6/21/88   REACTOR COOLANT LOOP "C" COLD LEG RESISTANCE TEMPERATURE DETECTOR (RTD) WAS READING LOW. FOUND BY OPERATOR WITH UNIT AT POWER. DEFECTIVE RTD READING 4 DEGREES LOW, ROOT CAUSE UNKNOWN. RTD TO BE REPLACED DURING UNIT REFUELING OUTAGE. (WR 135046)

10/17/88   DURING UNIT OPERATION, THE FIRE HAZARDS PANEL INDICATOR FOR THE 2C WIDE RANGE COLD LEG FAILED HIGH. THIS WAS DUE TO FAILURE OF THE 2C COLD LEG RTD (RESISTANCE TEMPERATURE DETECTOR). THE RTD WAS FOUND TO BE DEFECTIVE. THE RTD WAS REPLACED AND TESTED (B61418).

TABLE 8

TYPICAL EXAMPLES OF FAILURES DUE TO SHORT CIRCUIT

Date of Event	Failure Narrative
1/6/84	WITH THE UNIT AT POWER OPERATION, PERSONNEL PERFORMING A PERIODIC TEST ON THE REACTOR COOLANT "C" LOOP COLD LEG (SPARE) NARROW RESISTANCE TEMPERATURE DETECTOR (RTD) DISCOVERED THAT IT HAD SHORTED. THIS FAILURE HAD NO SIGNIFICANT EFFECT ON SYSTEM OPERATION BECAUSE ONLY THE SPARE RTD FAILED. THE EXACT CAUSE OF FAILURE WAS UNKNOWN, BUT IT WAS PROBABLY DUE TO NORMAL/ABNORMAL WEAR OF THE RTD CAUSING LOW INSULATION RESISTANCE. THE RTD WAS REPLACED, LEAK CHECKED, AND CALIBRATED. IT WAS THEN RETURNED TO NORMAL SERVICE. (84-343)
2/8/84	AT HOT STANDBY, THE AUXILIARY LOOP 1 TAVE CONTROL INDICATOR FAILED ITS MONTHLY CHANNEL CHECK. THE LOOP 1 DELTA T/TAVE CONTROL HOT LEG RTD WAS FOUND SHORTED. THE CAUSE OF THE FAILED CHANNEL CHECK WAS FOUND TO BE A SHORTED DELTA T/TAVE CONTROL LOOP 1 HOT LEG RTD. THE DELTA T/TAVE CONTROL DEFEAT SWITCHES WERE PLACED TO DEFEAT LOOP 1. REPLACED RTD DURING REFUELING OUTAGE. RECALIBRATED AND RETURNED TO SERVICE SAT.
4/26/84	WHILE AN OPERATOR WAS TAKING ROUTINE READINGS FROM THE QSPDS, HE NOTED THAT THE 1B2 T-COLD READ 25 DEGREES F LOWER THAN THE OTHER CHANNELS OF TC. THE PLANT WAS SHUT DOWN FOR REFUELING AND THIS HAD NO SIGNIFICANT EFFECT ON PLANT STATUS. IT WAS DETERMINED THAT THE TEMPERATURE ELEMENT FOR THE 1B2 T COLD TEMPERATURE LOOP HAD A SHORTED COMPENSATOR LEG. TE-1122CB WAS REPLACED WITH A NEW TE FROM STORES. PERFORMED LOOP CALIBRATION AS PER I & C PROCEDURE NO. 1-1400153R-11. RETURNED TO SERVICE. PWO NO. 6416
6/14/84	WHILE THE UNIT WAS AT FULL POWER IT WAS DISCOVERED THAT THE REACTOR COOLANT SYSTEM (RCS) RESISTANCE TEMPERATURE DETECTOR (RTD) FOR THE "B" CHANNEL HOT LEG INPUT TO THE REACTOR PROTECTION SYSTEM (RPS) (TE-1122HB) HAD A SHORTED LEAD. THIS HAD NO IMMEDIATE EFFECT ON NORMAL PLANT OPERATIONS. IT HAS NOT BEEN DETERMINED WHY THE LEADS SHORTED. A NEW RTD WAS CALIBRATED AND INSTALLED AND TE-1122HB WAS THEN RETURNED TO SERVICE. HAD TO WAIT UNTIL THE NEXT OUTAGE BEFORE THE RTD COULD BE CHANGED OUT. PWO NO. 6631
10/22/84	WITH THE UNIT IN POWER OPERATION, INSTRUMENTATION TECHNICIANS PERFORMING A LOOP NO. 1 DELTA T/TAVG CALCULATION TEST FOUND THAT THE SPARE RESISTANCE TEMPERATURE DETECTOR (RTD) 1RCTE1412C FAILED THE TIME RESPONSE PORTION OF THE TEST. NO SYSTEM FUNCTION WAS IMPAIRED SINCE THE FAILED RTD WAS A SPARE. THE FAILURE WAS CAUSED BY A SHORTED/GROUNDED CIRCUIT. THE FAILURE WAS CORRECTED BY INSTALLING AND TESTING A NEW SPARE RTD IN ACCORDANCE WITH PROCEDURE IMP-C-PROC-09 (REPLACEMENT OF A NARROW RANGE REACTOR COOLANT SYSTEM RTD). 86-097
11/06/85	WITH UNIT 2 IN COLD SHUTDOWN DURING ROUTINE CHECKS ASSOCIATED WITH START-UP, THE LOOP 'C' COLD LEG PROTECTION RESISTANCE TEMPERATURE DETECTOR TRANSMITTER TE-2-432C WAS FOUND TO BE GROUNDED. OPEN RTD DUE TO MANUFACTURING DEFECT. REPLACED RTD.
11/28/85	WITH UNIT IN REFUELING OPERATION, INSTRUMENTATION TECHNICIANS WERE PERFORMING A PERIODIC TEST (INSTRUMENTATION CALIBRATION PROCEDURE T-411 DELTA T/TAVG CALCULATION) WHEN THEY OBTAINED AN INCORRECT READING. SUBSEQUENT INVESTIGATION REVEALED THAT LOOP 'A' COLD LEG RESISTANCE TEMPERATURE DETECTOR (RTD) HAD FAILED. THE SYSTEM WAS NOT ADVERSELY AFFECTED BY THIS FAILURE SINCE THE UNIT WAS IN A REFUELING MODE OF OPERATION. THE CAUSE OF THE FAILURE IS UNKNOWN BUT IS BELIEVED TO HAVE RESULTED FROM A BREAK IN THE FLEXIBLE CONDUIT AT THE RTD CONNECTOR. THIS BREAK MAY HAVE BEEN THE PRODUCT OF ACTIVITY IN THE AREA DURING THE REFUELING OUTAGE. THE

FAILURE WAS CORRECTED BY INSTALLING AND TESTING A NEW RTD. AFTER THE TESTS WERE SATISFACTORY THE SYSTEM WAS RETURNED TO SERVICE. 86-099

12/24/85   WITH THE UNIT IN START-UP, SURVEILLANCE TESTING REVEALED A FAILURE ON THE LOOP 'C' T HOT SPARE RESISTANCE TEMPERATURE DETECTOR (RTD). THIS FAILURE HAD NO SIGNIFICANT EFFECT ON SYSTEM OPERATIONS BECAUSE IT WAS A SPARE. THE EXACT CAUSE OF THE FAILURE IS UNKNOWN, BUT IS BELIEVED TO BE DUE TO A SHORTED/GROUNDED CIRCUIT. THE FAILURE WAS CORRECTED BY INSTALLING A NEW RTD. AFTER SATISFACTORY TESTING, THE RTD WAS RETURNED TO SPARE STATUS. 86-385

12/7/85   THE CONTROL ROOM REPORTED THAT ONE TEMPERATURE INPUT TO THE REACTOR PROTECTION SYSTEM FOR THE 1B1 COLD LEG TEMPERATURE WAS READING LOW. THE UNIT WAS AT FULL POWER AND THIS HAD NO IMMEDIATE EFFECT ON NORMAL PLANT OPERATIONS. IT WAS DETERMINED THAT TE-1122CA WAS INTERNALLY SHORTED. THE RESISTANCE TEMPERATURE DETECTOR (TE-1122CA) WAS CHANGED OUT WITH A NEW ONE FROM STORES AND TE-1122CA WAS RETURNED TO SERVICE. HAD TO WAIT UNTIL THE NEXT REFUELING OUTAGE. PWO NO. 8076

7/24/86   DURING FULL POWER OPERATION, OPERATORS NOTED A SPURIOUS TEMPERATURE INDICATION FROM 'A' LOOP BYPASS MANIFOLD CHANNEL TE-401A. LATER, THE CHANNEL INITIATED AN OVERPOWER CHANGE IN TEMPERATURE TURBINE RUNBACK ALARM. INSTRUMENT AND CONTROL PERSONNEL DISCOVERED RESISTANCE TEMPERATURE DETECTOR 2-TE-401A HAD A LOW SIGNAL WIRE TO GROUND RESISTANCE. CURRENTLY IT IS UNKNOWN WHY THE DETECTOR FAILED. AN INSTALLED SPARE DETECTOR TE-405A, WAS SUBSTITUTED AS THE SUPPLY TO CHANNEL TE-401A. THE CHANNEL WAS CALIBRATED AND RESTORED TO SERVICE.

6/16/87   REACTOR COOLANT SYSTEM LOOP C DIFFERENTIAL TEMPERATURE AVERAGE PROTECTION RESISTANCE TEMPERATURE DETECTOR TE-3-432D FAILED SPECIAL TEST. UNIT AT REFUELING SHUTDOWN CONDITIONS. TE-3-432D SHORTED TO GROUND. IT IS BELIEVED THAT THE FAULT WAS CAUSED BY IMPROPER CONSTRUCTION WORK PRACTICES IN THE AREA. TE-3-432D REPLACED. TESTED SATISFACTORILY. *8985-63*2

8/25/87   NTP-140 LOOP 4 REACTOR COOLANT SYSTEM AVERAGE TEMPERATURE, DIFFERENTIAL TEMPERATURE, AND OVER TEMPERATURE SETPOINT WERE OBSERVED TO BE SWINGING ERRATICALLY. (JO-013253) FOUND NOISE SUPPRESSOR AT RESISTANCE TEMPERATURE DETECTOR TERMINAL BLOCK TO BE SHORTED. REPLACED NOISE SUPPRESSOR. MONITORED OPERATION OF CIRCUIT FOR TWENTY FOUR HOURS AND FOUND CIRCUIT TO BE FUNCTIONING PROPERLY.

12/8/87   UNIT AT FULL POWER. REACTOR PROTECTION SYSTEM (RPS) TEMPERATURE INDICATION WAS READING HIGH. NO EFFECT ON PLANT. TEMPERATURE ELEMENT TE-1112HC FAILED (SHORTED TO GROUND), CAUSE UNKNOWN. LEADS LIFTED TO CLEAR GROUND, TE-1112HC WILL BE REPLACED AT THE NEXT OUTAGE. PWO NO. 7389 PWO NO. 7478 PWO NO. 7466

12/30/87   WHILE THE PLANT WAS AT FULL POWER THE CONTROL ROOM REPORTED THAT THE TEMPERATURE ELEMENT FOR REACTOR COOLANT LOOP 1A2 COLD LEG TE-1112CC WAS CAUSING THE THERMAL MARGIN/LOW PRESSURE (TM/LP) TRIP FOR THE REACTOR PROTECTION SYSTEM (RPS) TO TRIP. THIS HAD NO IMMEDIATE EFFECT ON NORMAL PLANT OPERATIONS. IT WAS DETERMINED THAT TE-1112CC WAS GROUNDED SO THE NECESSARY LEADS WERE LIFTED TO CLEAR THE RPS OF TM/LP TRIPS. THE REASON FOR THE GROUNDING IS UNKNOWN. TE-1112CC WILL BE PERMANENTLY REPAIRED DURING THE NEXT OUTAGE. PWO NO. 6807

11/9/88   THE CONTROL ROOM REPORTED THAT TE-1122HC (TEMPERATURE ELEMENT FOR REACTOR COOLANT LOOP 2B HOT LEG) WAS CAUSING SPIKES ON THE RPS (REACTOR PROTECTION SYSTEM) WHICH WAS CAUSING SEVERAL RPS CHANNEL 'C' TRIPS. TRIPS MUST BE BYPASSED THEREFORE CHANGING THE TRIP LOGIC. IT WAS DETERMINED THAT THE TEMPERATURE ELEMENT LEAD WAS CRIMPED UNDER THE LEAD BOX COVER AND WAS BEING SHORTED TO GROUND. IMPROPER INSTALLATION. THE LEADS WERE REARRANGED SO THEY WOULD NOT BE CRIMPED AND THE CRIMPED LEAD WAS REPAIRED AND TE-1122HC WAS RETURNED TO NORMAL. PWO NO. 6882 TKM

TABLE 9

TYPICAL EXAMPLES OF FAILURES FROM DEFECTIVE CONNECTION

Date of Event	Failure Narrative
7/2/82	THE PRIMARY COOLANT LOOP 1A COLD LEG TEMPERATURE SENSOR RESISTANCE TEMPERATURE DETECTOR (RTD) WAS FOUND TO BE LOW OUT OF SPECIFICATION ON TEMPERATURE INDICATION. A LOOSE AND OXIDIZED CONNECTION CAUSED THE LOWER READING. TIGHTENED AND CLEANED THE CONNECTION.
3/22/83	DURING THE REACTOR COOLANT SYSTEM RESISTANCE TEMPERATURE DETECTOR (RTD) SURVEILLANCE, CHANNEL B PRIMARY COOLANT TEMPERATURE MONITOR TE-112-HB WAS OUT OF TOLERANCE AND THE SPREAD BETWEEN TE-112-CA, CB, CC, CD WAS ALSO OUT OF SPEC. THE SPREAD BETWEEN RTD'S WAS ATTRIBUTED TO INSTRUMENT DRIFT AND OXIDATION OF CONNECTIONS. TE-112-HB WAS OUT OF TOLERANCE DUE TO A FAULTY CONNECTION. RECALIBRATED AND CLEANED THE RESPECTIVE COMPONENTS.
1/29/86	WITH THE UNIT AT POWER OPERATION, SHIFT PERSONNEL OBSERVED ERRATIC INDICATION ON "B" LOOP DELTA T/TAVG CONTROL. THIS RESULTED IN NO SIGNIFICANT EFFECT SINCE THE CHANNEL COULD BE BYPASSED. THE CAUSE OF THE FAILURE WAS A DEFECTIVE HOT LEG RESISTANCE TEMPERATURE DETECTOR (RTD), PROBABLY CAUSED BY DEFECTIVE CONNECTION OR LOOSE PARTS. THE RTD WAS REPLACED AND THE CHANNEL RECALIBRATED SATISFACTORILY. 86-603
2/4/86	DURING NORMAL OPERATION, CONTROL ROOM PERSONNEL NOTED THAT THE "A" MAIN STEAMLINE TUNNEL TEMPERATURE INSTRUMENT WAS INOPERABLE, READING AT LEAST 20 DEGREES F. ABOVE THE THREE REMAINING INSTRUMENTS. THERE WAS NO SIGNIFICANT EFFECT ON SYSTEM OPERATION. REF. MWO 2-86-0630. THE FAILURE WAS ATTRIBUTED TO THE "A" RESISTANCE TEMPERATURE DETECTOR (RTD) WIRING BEING FOUND LOOSE AT ITS CONNECTION POINTS TO THE TERMINAL STRIP IN THE CONTROL ROOM PANEL. INADEQUATE REPAIR OF THIS COMPONENT IS CONSIDERED TO BE THE CAUSE OF FAILURE. THE LOOSE RTD WIRING WAS TIGHTENED AT THE TERMINAL STRIP CONNECTIONS. THE TEMPERATURE INSTRUMENT READING WAS THEN FOUND TO BE COMPARABLE TO THE REMAINING INSTRUMENTS AND THE CHANNEL WAS DECLARED OPERABLE.
6/3/86	WITH UNIT AT 0% POWER, ENGINEERING NOTED THE LOOP 1 HOT LEG TEMPERATURE INDICATION WAS SIGNIFICANTLY OUTSIDE CORE PROTECTION CALCULATOR ACCEPTANCE TOLERANCES. INVESTIGATION OBSERVED A 5 DEGREE DIFFERENCE ON RESISTANCE TEMPERATURE DETECTOR (RTD) AND AT PLANT PROTECTION SYSTEM (PPS) LOOPS. ENTERED REACTOR CONTAINMENT AND FOUND RTD WITH A LOOSE CONNECTION. REPAIRED LOOSE CONNECTION AND PERFORMED A 1-POINT CALIBRATION. VERIFIED READINGS AND RETURNED TO SERVICE SATISFACTORILY.
10/21/87	WHILE THE UNIT WAS SHUT DOWN FOR REPAIRS, IT WAS REPORTED BY SURVEILLANCE PERSONNEL THAT THE LEADS ON A REACTOR COOLANT LOOP "B" COLD LEG TEMPERATURE ELEMENT WERE READING A HIGHER RESISTANCE THAN DESIRED. THE LEADS ON THE TERMINAL BLOCK HAD BECOME LOOSE RESULTING IN THE IMPROPER READINGS. THE ROOT CAUSE IS UNKNOWN. CLEANED AND TIGHTENED THE LEADS ON THE TERMINAL BLOCK. RETURNED TO SERVICE AND VERIFIED PROPER OPERATION.
12/23/87	REACTOR COOLANT LOOP 1A TO CHANNEL 'A' REACTOR PROTECTION SYSTEM TEMPERATURE INDICATOR WAS PEGGED LOW. POOR INDICATION WAS CAUSED BY BAD CONNECTION OF RESISTANCE TEMPERATURE DETECTOR LEADS AT THE TRANSMITTER. THE RESISTANCE TEMPERATURE DETECTOR LEADS WERE PULLED AND RECONNECTED, WHICH RESTORED INDICATION.

TABLE 10

EXAMPLES OF FAILURES FROM AGING/CYCLIC FATIGUE

Date of Event	Failure Narrative
1/30/79	CHANNEL C PRIMARY COOLANT RESISTANCE TEMPERATURE DETECTOR (RTD) TE-112-HC RESPONSE TIME TESTING EXCEEDED SPECIFICATIONS. UNKNOWN. ELIMINATED TE-112-HC AS A REACTOR PROTECTION SYSTEM INPUT.
3/29/83	DURING A RESISTANCE TEMPERATURE DETECTOR RESPONSE TIME SURVEILLANCE, A RESISTANCE TEMPERATURE DETECTOR (RTD) FAILED TO MEET THE RESPONSE TIME CRITERIA. UNKNOWN. THE RTD IS INACCESSIBLE DURING POWER OPERATION. THE RTD WAS REMOVED FROM THE REACTOR PROTECTION SYSTEM VIA THE BYPASS/JUMPER SYSTEM.
1/6/84	WITH UNIT AT POWER, OPERATIONS PERSONNEL RECEIVED AN ALARM INDICATING THAT THE REACTOR COOLANT SYSTEM LOOP "B" NARROW RANGE TEMPERATURE LOOP HAD FAILED DOWNSCALE. THE I&C DEPARTMENT DETERMINED THAT THE RESISTANCE TEMPERATURE DETECTOR (RTD) HAD FAILED OPEN. THE CAUSE OF FAILURE WAS UNKNOWN BUT COULD HAVE BEEN THE RESULT OF AGING CYCLIC FATIGUE. THE RTD WAS REPLACED. (84-278)
2/9/84	WITH UNIT SHUTDOWN, I&C PERSONNEL PERFORMING TIME RESPONSE TESTING ON THE WIDE RANGE T HOT RTD LOOP DISCOVERED THAT THE RTD WAS BREAKING DOWN DURING TESTING AND WAS OPERABLE BUT ABOUT TO FAIL. THERE WAS NO SIGNIFICANT EFFECT BECAUSE THE RTD WAS REPLACED BEFORE IT DID ACTUALLY FAIL. THE CAUSE OF FAILURE WAS UNKNOWN, BUT COULD HAVE BEEN THE RESULT OF AGING/CYCLIC FATIGUE. THE RTD WAS REPLACED AND THE LOOP RECALIBRATED AND RETURNED TO SERVICE PER STATION PROCEDURE. 84-276
9/27/84	WITH THE UNIT IN POWER OPERATION, OPERATIONS SHIFT PERSONNEL DETECTED A FAILURE ON "C" REACTOR COOLANT LOOP TEMPERATURE INSTRUMENTATION. INVESTIGATION REVEALED THAT THE "C" CONTROL LOOP TC COLD LEG RESISTANCE TEMPERATURE DETECTOR (RTD) HAD FAILED. THIS FAILURE DID NOT AFFECT THE SYSTEM OPERATION SINCE THE SPARE RTD WAS BROUGHT INTO SERVICE. THE CAUSE OF THE FAILURE WAS AN OPEN CIRCUIT ON THE "C" LOOP TC COLD LEG RTD WHICH RESULTED FROM NORMAL WEAR AND AGING. THE FAILURE WAS CORRECTED BY INSTALLING AND TESTING A NEW RTD IN ACCORDANCE WITH THE REPLACEMENT OF A NARROW RANGE REACTOR COOLANT RTD PROCEDURE AND THE INSTRUMENT CALIBRATION PROCEDURE FOR DELTA T/TAVG CALCULATION. THE SPARE RTD WAS RETURNED TO SPARE STATUS. 86-100
3/13/85	WITH UNIT IS STARTUP MODE, OPERATIONS IDENTIFIED LEAKAGE ON LOOP 1 HOT LEG TEMPERATURE THERMOWELL. AN UNUSUAL EVENT WAS DECLARED AND THE UNIT BROUGHT OFF LINE TO REPAIR. INVESTIGATION FOUND WATER IN THERMOWELL THAT CAUSED FAILURE OF THE THERMOCOUPLE. WATER WAS FROM A BLOWN SPRAY VALVE PACKING IN NEARBY AREA. REFERENCE: NCR3-1038, NCR3-1039. MOVED TEMPERATURE SENSOR TO ADJACENT THERMOWELL. PLUGGED OLD THERMOWELL UNTIL REFUELING OUTAGE. OLD THERMOWELL REMOVED WAS TESTED FOR LEAKS AND NONE WAS FOUND. ANALYSIS NOTED WATER CONTENT SIMILAR TO WATER FROM A VALVE PACKING LEAK. REPAIRED WITH NEW THERMOWELL AND RETURNED TO SERVICE.
12/10/85	WHILE THE UNIT WAS SHUT DOWN FOR REFUELING, THE 18 MONTH RESISTANCE TEMPERATURE DETECTOR (RTD) TIME RESPONSE SURVEILLANCE WAS BEING PERFORMED

AND RTD 1112HB (TE-1112HB) DID NOT PASS. THIS HAD NO IMMEDIATE EFFECT ON NORMAL PLANT OPERATIONS. THE CAUSE OF FAILURE IS UNKNOWN BUT IS BELIEVED TO BE END OF LIFE. A NEW RTD WAS CALIBRATED, INSTALLED AND TE-1122HB WAS RETURNED TO SERVICE. HAD TO WAIT UNTIL THE REFUELING OUTAGE. PWO NO. 8076

12/7/85   WHILE THE UNIT WAS SHUT DOWN FOR REFUELING AND THE 18 MONTH RESISTANCE TEMPERATURE DETECTOR (RTD) RESPONSE TIME SURVEILLANCE WAS BEING PERFORMED AND RTD 1122HA (TE-1122HA) DID NOT PASS. THIS HAD NO IMMEDIATE EFFECT ON NORMAL PLANT OPERATIONS. THE CAUSE OF FAILURE IS UNKNOWN BUT IS BELIEVED TO BE END OF LIFE. A NEW RTD WAS CALIBRATED, INSTALLED AND TE-1122HA WAS RETURNED TO SERVICE. HAD TO WAIT UNTIL THE NEXT REFUELING OUTAGE. PWO NO. 8076, PWO NO. 8029, PWO NO. 6654

4/17/86   WITH THE UNIT IN POWER OPERATION, OPERATIONS PERSONNEL DETECTED A FAILURE OF THE A' LOOP DELTA TEMPERATURE CONTROL. THE REACTOR TEMPERATURE DETECTOR (RTD) WAS DRIFTING LOW OUT OF SPECIFICATION. THERE WAS NO SIGNIFICANT EFFECT ON PLANT OPERATION BECAUSE INDICATION IN THE CONTROL ROOM NOTED THIS FAILURE ALLOWING A BYPASS OF THE RTD. THE OTHER TWO LOOPS INDICATED NORMALLY. THE CAUSE OF THE FAILURE WAS BELIEVED TO BE A RESULT OF COMPONENT AGING. AS A CORRECTIVE ACTION THE INSTALLED SPARE WAS JUMPERED INTO SERVICE UNTIL THE RTD COULD BE REPLACED. 86-305

5/1/86   WITH THE UNIT IN POWER OPERATIONS, SHIFT PERSONNEL DETECTED A FAILURE ON TH 'C' LOOP DELTA T/TAVG PROTECTION INSTRUMENTATION. INVESTIGATION REVEALED A FAILED RESISTANCE TEMPERATURE DETECTOR (RTD). THIS FAILURE RESULTED IN THE LOSS OF A SYSTEM FUNCTIONAL PATH. THE EXACT CAUSE OF THE FAILURE IS UNKNOWN. THE OPEN CIRCUIT IS BELIEVED TO BE THE RESULT OF COMPONENT AGING. THE FAILURE WAS CORRECTED BY REPLACING THE FAILED RTD, CALIBRATING AND TESTING THE NEW RTD, AND RETURNING THE CIRCUIT TO SERVICE. 86-573

7/20/86   WITH UNIT 1 AT 54% POWER, CONTROL ROOM INDICATIONS SHOWED GREATER THAN 6% DIFFERENCE IN CHANNELS OF THE 'B' LOOP DELTA T PROTECTION TRANSMITTER (TE-1-422A). (86-063. 4C1 WO: 038745) FAILURE OF RESISTANCE TEMPERATURE DETECTOR (RTD) DUE TO AGE. REPLACED RTD.

8/6/86   WITH THE UNIT IN HOT SHUTDOWN, SHIFT PERSONNEL DETECTED A FAILURE ON THE 'A' LOOP DELTA T/TAVG PROTECTION INSTRUMENTATION. INVESTIGATION REVEALED A FAILED RESISTANCE TEMPERATURE DETECTOR (RTD). THIS FAILURE RESULTED IN THE LOSS OF A SYSTEM FUNCTIONAL PATH. THE EXACT CAUSE OF THE FAILURE IS UNKNOWN. THE OPEN CIRCUIT IS BELIEVED TO BE THE RESULT OF COMPONENT AGING. THE FAILURE WAS CORRECTED BY REPLACING THE FAILED RTD, CALIBRATING AND TESTING THE NEW RTD, AND RETURNING THE CIRCUIT TO SERVICE. 86-534

8/11/86   WITH THE UNIT IN MODE 5, I&C TECHNICIANS FOUND THE RESISTANCE TEMPERATURE DETECTOR (RTD) FOR TEMPERATURE (T) HOT CHANNEL II DELTA T/T AVG SPIKING INTERMITTENTLY. THERE WAS NO SIGNIFICANT AFFECT ON PLANT OPERATION BECAUSE DELTA T AND T AVG ARE AUCTIONEERED HIGH AND THE OTHER TWO CHANNELS WERE OPERATING PROPERLY. THE CAUSE OF THE FAILURE WAS DUE TO FAILED RTD, PROBABLY CAUSED BY WEAR/AGING. AS A CORRECTIVE ACTION THE RTD WAS REPLACED. 86-605

3/24/87   WITH THE UNIT AT 100% POWER, OPERATIONS STATED THAT THE LOOP 1 HOT LEG TEMPERATURE INSTRUMENT INDICATIONS WERE ERRATIC AND CAUSING DEPARTURE FROM NUCLEATE BOILING RATIO (DNBR) TRIPS OF THE CORE PROTECTION CALCULATOR (CPC) CHANNEL 'A'. INVESTIGATION AND TROUBLESHOOTING FOUND THE TEMPERATURE ELEMENT WAS BAD. SUSPECT PREMATURE END OF LIFE. CHANGED OVER TO THE SPARE TEMPERATURE ELEMENT BY LIFTING WIRES. CALIBRATED AND RETURNED TO SERVICE SATISFACTORILY. DURING THE NEXT OUTAGE THE ELEMENT WILL BE REPLACED.

| 11/4/87 | THE COMPUTER POINT A1692 FOR THE REACTOR PROTECTION SYSTEM CHANNEL "A" HOT LEG TEMPERATURE INDICATION WAS READING AN OPEN CONDITION. THE TECHNICIANS FOUND THAT RESISTANCE TEMPERATURE DETECTOR 1RC_RD0001A, USED TO MEASURE THE HOT LEG TEMPERATURE, HAD AN OPEN CIRCUIT. WE SUSPECT THAT AGING CONTRIBUTED TO THE COMPONENT FAILURE. THE TEMPERATURE DETECTOR WAS REPLACED AND ITS OPERATION WAS VERIFIED TO THE SPECIFICATIONS OF THE CONTROLLING PROCEDURE. THE COMPUTER READING RETURNED TO NORMAL. |

| 1/27/87 | WITH UNIT AT COLD SHUTDOWN, PERSONNEL DISCOVERED DURING CALIBRATION TESTING THAT THE LOOP "A" HOT CONTROL RESISTANCE TEMPERATURE DETECTOR (TE-1-411B) HAD AN INSULATION BREAKDOWN AND WAS SENDING INCORRECT SIGNALS. (87-030. 4C1 WO: 048806) INSULATION BREAKDOWN DUE TO HEAT. INSTALLED NEW RESISTANCE TEMPERATURE DETECTOR. |

| 3/11/87 | WITH UNIT AT POWER, ENGINEERING PERSONNEL PERFORMING RESISTANCE TEMPERATURE DETECTOR (RTD) CROSS CALIBRATION PROCEDURES CONCLUDED THAT THE RTD THAT THIS LOOP WAS TEMPORARILY CONNECTED TO (AN INSTALLED SPARE) HAD DRIFTED TEN DEGREES OUT OF SPECIFICATION. THE CHANNEL WAS PLACED IN DEFEAT UNTIL THE RTD COULD BE REPLACED. THE CAUSE OF FAILURE WAS ASSUMED TO BE AGING/CYCLIC FATIGUE OF THE RTD. THE RTD WAS REPLACED AT THE FIRST OUTAGE AVAILABLE AND THE LOOP WAS SWITCHED BACK TO THE NORMAL RTD. 87-275 |

| 5/19/87 | WITH UNIT 1 IN COLD SHUTDOWN DURING SURVEILLANCE TESTING, PERSONNEL FOUND THAT THE LOOP "A" SPARE RESISTANCE TEMPERATURE DETECTOR (RTD) (TE-1-411D) WAS READING OPEN INSTEAD OF CLOSED. (LASKOWSKI, WO: 053478, 87-081. 4C1) WEAR AND AGING ATTRIBUTED FROM SYSTEM STRESS. REPLACED RTD. |

| 7/13/87 | UNIT 2 WAS OPERATING AT 100% POWER. CONTROL ROOM OPERATORS OBSERVED ERRATIC OUTPUT ON THE REACTOR COOLANT SYSTEM HOT LEG TEMPERATURE INDICATOR (TE-2-433). (BASS, WO: 055075, 88-193. 4C2) EXACT CAUSE WAS UNKNOWN, BUT SUSPECT NORMAL WEAR AND AGING OF TEMPERATURE TRANSMITTER (TE-2-433) REPLACED TRANSMITTER LIKE FOR LIKE. |

| 8/7/87 | WITH THE UNIT AT 92% POWER, OPERATIONS OBSERVED THE REACTOR COOLANT COLD LEG TEMPERATURE LOOP 'C' TO BE INDICATING 10 DEGREES LESS THAN NORMAL. READINGS WERE 510 TO 520 DEGREES INSTEAD OF THE NORMAL 520 TO 530 DEGREES. INVESTIGATION CONCLUDED THAT THE TEMPERATURE ELEMENT WAS FAILING AND THE RECORDER WAS FAILING. NORMAL WEAROUT/AGING OF THE TEMPERATURE ELEMENT. REFERENCE: NCR1-P-6209. REPLACED THE TEMPERATURE ELEMENT WITH A NEW ONE. TESTED AND RETURNED TO SERVICE. THE RECORDER WILL BE REPLACED AT A LATER DATE. |

| 10/31/87 | WITH THE UNIT IN HOT SHUTDOWN, OPERATORS OBSERVED THAT THE "B" LOOP DELTA T/TAVG PROTECTION CHANNEL HAD FAILED HIGH. UNIT OPERATION WAS NOT AFFECTED AS THE UNIT WAS ALREADY TRIPPED. THE RESISTANCE TEMPERATURE DETECTOR (RTD) THAT FEEDS THE "B" LOOP PROTECTION CHANNEL FAILED DUE TO AGING/CYCLIC FATIGUE OR ABNORMAL STRESSES DUE TO VIBRATIONS ASSOCIATED WITH REACTOR COOLANT FLOW. THE SPARE RTD WAS CONNECTED TO THE "B" LOOP PROTECTION CHANNEL, CALIBRATED SATISFACTORILY AND THE CHANNEL RETURNED TO SERVICE. THE FAILED RTD WILL BE REPLACED. 88-093 |

| 2/7/88 | WITH THE UNIT IN HOT SHUTDOWN, INSTRUMENT TECHNICIANS WERE PERFORMING A CALIBRATION PROCEDURE ON THE "B" REACTOR COOLANT LOOP RESISTANCE TEMPERATURE DETECTOR (RTD) AND FOUND AN OPEN CIRCUIT ACROSS THE RTD. NO EFFECT ON THE SYSTEM OPERATION AS RTD SUPPLIER INPUT TO OVERPOWER TEMPERATURE PROTECTION WHICH IS NOT REQUIRED IN HOT SHUTDOWN. A BROKEN WIRE IN THE RTD CIRCUIT WOULD CAUSE AN OPEN CIRCUIT. THE MOST LIKELY CAUSE |

WAS ABNORMAL STRESSES DUE TO VIBRATION OR AGING FATIGUE. A SPARE RTD WAS CONNECTED TO SUPPLY AN INPUT TO THE OVERPOWER PROTECTION CIRCUIT. THE CIRCUIT WAS CALIBRATED SATISFACTORILY AND RETURNED TO SERVICE PRIOR TO POWER OPERATIONS. 88-064

2/21/88   WITH THE UNIT AT POWER, CONTROL OPERATORS OBSERVED THE "B" LOOP HOT LEG TEMPERATURE FAILED LOW CAUSING A LOW DIFFERENTIAL TEMPERATURE (DELTA "T") AND LOW AVERAGE TEMPERATURE ("TAVG") ALARM TO BE ANNUNCIATED. THE CHANNEL WAS PLACED IN TRIP TO ENSURE REACTOR PROTECTION AND NO OTHER EFFECT ON PLANT OPERATION OCCURRED. COMPENSATION LEAD ON RESISTANCE TEMPERATURE DETECTOR (RTD) OPENED CAUSING RTD TO FAIL LOW. MOST LIKELY CAUSE WAS ABNORMAL STRESSES OF FLUID VIBRATION OR AGING/CYCLIC FATIGUE. SPARE COMPENSATION LEAD ON THE RTD WAS CONNECTED. THE RTD WAS RECALIBRATED AND THE CHANNEL RETURNED TO SERVICE. (88-058)

3/23/88   WITH UNIT 2 AT 100% POWER, CONTROL ROOM OPERATORS RECEIVED INDICATIONS THAT THE LOOP 'A' REACTOR COOLANT SYSTEM WIDE RANGE COLD LEG TEMPERATURE TRANSMITTER (TE-2-410) WAS FAILING LOW. (BASS, WO: 062888, 88-052. 4C2) CAUSE DUE TO END OF LIFE OF SUPPRESSION BOARD TO THE LOW LEVEL AMPLIFIER. REPLACED SUPPRESSION BOARD.

APPENDIX D

ITS-90 SCALE

THE INTERNATIONAL TEMPERATURE SCALE OF 1990

I.   INTRODUCTION

On January 1, 1990 a new temperature scale called the International Temperature Scale of 1990 (ITS-90) was adopted and is now being implemented in the United States by the National Institute of Standards and Technology (NIST)[1]. The ITS-90 supersedes the previous temperature scale, the International Practical Temperature Scale of 1968 (IPTS-68). The ITS-90 also replaces the 1976 Provisional 0.5 Kelvin (K) to 30 K Temperature Scale (EPT-76).

The ITS-90 was developed because of a number of deficiencies and limitations of the IPTS-68. These include:

1)   The IPTS-68 had a lower temperature limit of -259.34°C.

2)   The IPTS-68 was inaccurate in reproducing thermodynamic temperatures especially in the range of 630.74°C to 1337.58°C.

The new temperature scale now extends down to -272.5°C. The inaccuracies of the IPTS-68 in the range of 630 to 1338°C were improved by replacing the standard interpolating instrument used in that range. In the IPTS-68, the standard platinum resistance thermometer (SPRT) was used as the standard interpolating instrument from -259.34 to 630.74°C. From 630.74°C to 1064.43°C, the standard interpolating instrument was the platinum -10% rhodium versus platinum thermocouple. The ITS-90 now uses the SPRT for the standard interpolating instrument from -259.3467 to 961.78°C. Temperatures below -259.3467°C are defined by vapor pressure relationships and the constant volume gas thermometers (CVGT). Temperatures above 961.78°C in the ITS-90 are defined by radiation pyrometry.

Figure 1 shows the interpolating instruments and their respective ranges for the IPTS-68 and ITS-90. The temperature range of interest for primary coolant temperature measurements in PWRs is 0 to 400°C. Although the interpolation instrument (SPRT) was not changed from

---

1.   B. W. Mangum, "A New Temperature Scale, The International Temperature Scale of 1990, Is Adopted", National Institute of Standards and Technology.

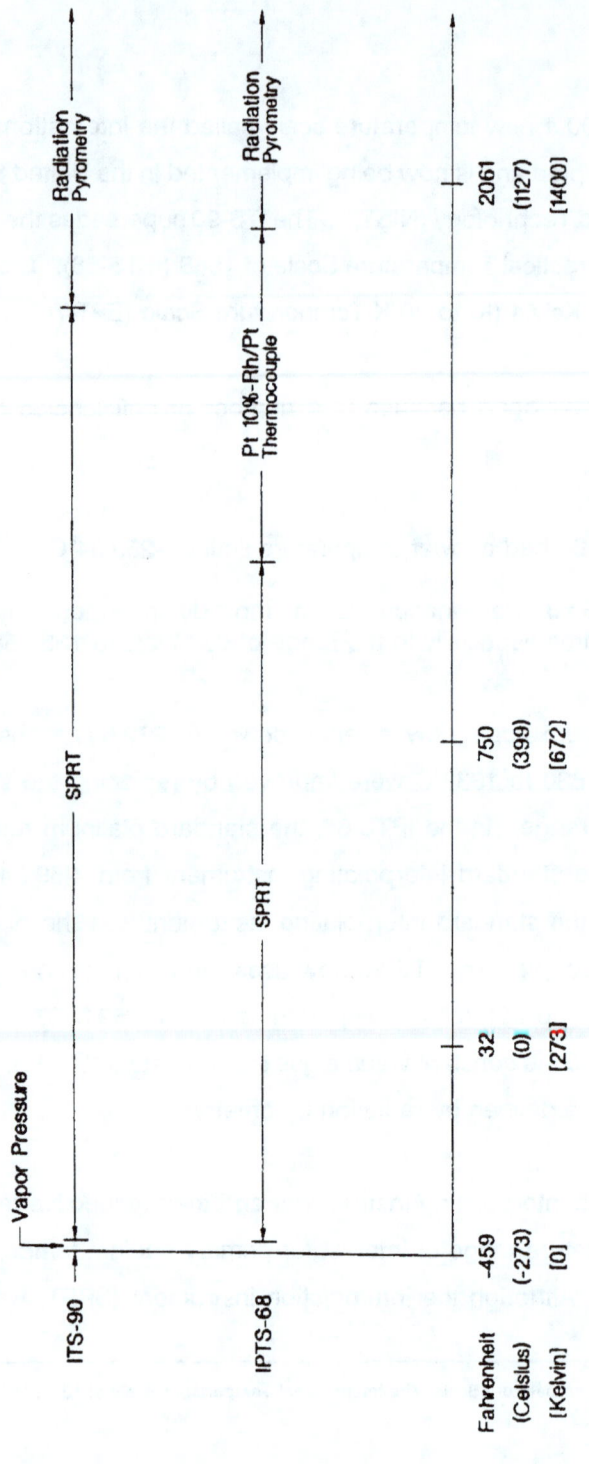

Figure 1. Interpolation Instruments for IPTS–68 and ITS–90.

IPTS-68 to ITS-90 in this range, changes in the defined temperatures of fixed points and the way in which the SPRT is calibrated and used for interpolation result in a maximum difference of nearly 0.05°C in the 0 to 400°C range. Consequently, there is an impact on temperature measurement for the nuclear power industry as a result of the implementation of ITS-90.

II.   ITS-90 USE OF AN SPRT

First a discussion of nomenclature is warranted. As in IPTS-68, the unit of thermodynamic temperature for the ITS-90 is the Kelvin (K). A Kelvin is defined as 1/273.16 of the thermodynamic temperature of the triple point (TP) of water. Temperatures expressed in Kelvin are represented by an upper case T. The celsius thermodynamic temperature, represented by a lower case t, is defined by:

$$t = T - 273.15 \tag{1}$$

Units for the celsius scale are °C. Thermodynamic temperatures (T or t) have no subscript. Practical temperatures are denoted with a subscript indicating which scale is being used as in T_{68} or t_{90}. The upper or lower case t still designates which units are being used.

The interpolating equations used with SPRTs are defined in terms of a resistance ratio (W). For IPTS-68 the resistance ratio of an SPRT was described relative to its ice point resistance $R(0.0°C)$:

$$W_{68}(T_{68}) = R(T_{68})/R(273.15 \ K) \tag{2}$$

or

$$W_{68}(t_{68}) = R(t_{68})/R(0.0°C) \tag{3}$$

For ITS-90 the resistance ratio of an SPRT is described relative to its resistance at the triple point of water.

$$W_{90}(T_{90}) = R(T_{90})/R(273.16 \ K) \tag{4}$$

or

$$W_{90}(t_{90}) = R(t_{90})/R(0.01°C) \tag{5}$$

According to the ITS-90, the SPRT is used for interpolating between the fixed points over the range of 13.8033 K (-259.3467°C) to 1234.93K (961.78°C). This range is broken down into two regions:

Range 1 - 13.8033 K to 273.16 K

Range 2 - 273.15 K to 1234.93 K

For each of the above ranges, ITS-90 defines a reference function ($W_r(T_{90})$) which describes a nominal SPRT ratio resistance versus temperature. For the range of 273.15 K to 1234.93 K a 9th order polynomial is used:

$$W_r(T_{90}) = C_0 + \sum_{i=1}^{9} C_i \left(\frac{T_{90} - 754.15}{481} \right)^i \qquad (6)$$

The constant (C_0) and the coefficients (C_i) are tabulated in ITS-90. It is important to note that this reference function is for a nominal SPRT and does not contain any information about the calibration of a particular SPRT. Figure 2 shows a plot of the reference function. To illustrate the curvature of the reference function, a straight line is also plotted between the end points of the reference function. The difference between the reference function and the straight line is also shown and labeled on the right hand y-axis.

There is a similar reference function to describe the nominal SPRT over the range of 13.8033 K to 273.16 K. It is a 12th order polynomial.

For each of the two ranges described by the two reference functions ($W_r(T_{90})$), the range is further divided into subranges. The range below 273.15 K is divided into four subranges while the range above 273.15 K is divided into six subranges. Figure 3 shows these ranges.

Subrange 5 overlaps 273.15 K from the triple point of mercury to the melting point of gallium. Note that there are no temperature ranges which rely on extrapolated temperature versus resistance behavior for the SPRT. This differs from the IPTS-68 range of 0 to 630.74°C

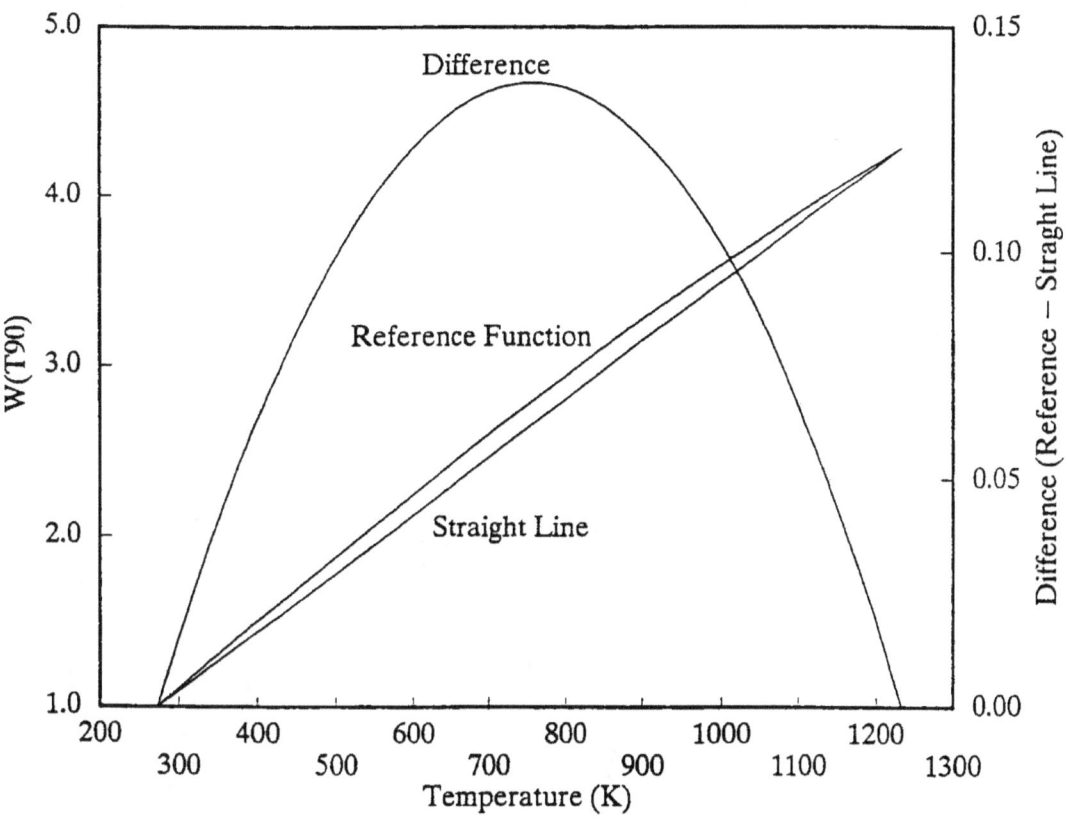

Figure 2. ITS-90 Reference Function and Difference Between the Reference Function and a Straight Line.

AMS—DWG CAL007A

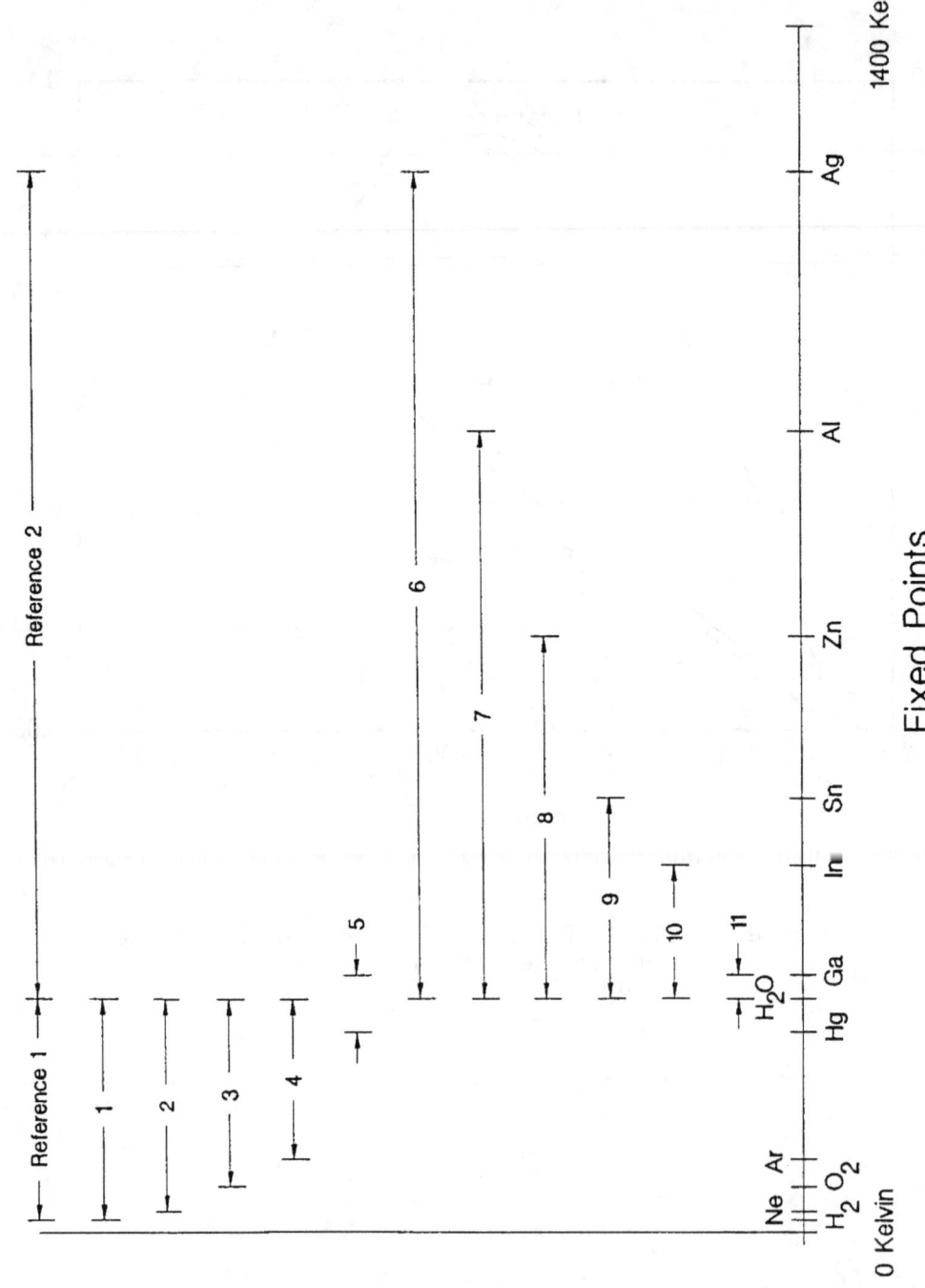

Figure 3. Reference Ranges and Subranges for SPRTs and Applied to ITS-90.

where temperatures from the zinc point (419.53°C) to 630.74°C were based on extrapolated resistance versus temperature data.

For each of the eleven subranges, there is an equation ($\Delta W_i(T_{90})$) to describe the deviation in resistance ratio of a given SPRT from the reference function. Thus, for the measured resistance ratio of an SPRT ($W(T_{90})$) the relation is as follows:

$$W(T_{90}) = W_r(T_{90}) + \Delta W_i(T_{90}) \tag{7}$$

For our application, subrange 8 is the appropriate subrange and its deviation function is:

$$\Delta W_8(T_{90}) = a_8[W(T_{90})-1] + b_8[W(T_{90})-1]^2 \tag{8}$$

For subrange 8, the SPRT must be calibrated at the triple point of water and the freezing points (FP) of tin and zinc. With these three measurements, $W(FP\ Sn)$ and $W(FP\ Zn)$ are calculated. Using Equation 6, values for $W_r(FP\ Sn)$ and $W_r(FP\ Zn)$ can be obtained and then with Equation 7, values for $\Delta W(FP\ Sn)$ and $\Delta W(FP\ Zn)$ can be calculated. Values for a_8 and b_8 are then obtained using Equation 8. These two values, a_8 and b_8, together with measurement of resistance at triple point of water will completely characterize an SPRT for the ITS-90 over subrange 8.

III. CONVERSION OF SPRT CALIBRATION TO ITS-90

An SPRT does not have to be recalibrated for the ITS-90. The data from the previous calibration to the IPTS-68 can be used to obtain an ITS-90 calibration. To provide an example, NIST calibration data of a sample SPRT (S/N 3447) was converted to an ITS-90 calibration. A step by step procedure of this conversion is given in this section to guide the reader. However, if the SPRT has not been calibrated at NIST for more than a year, it should be sent to NIST for a new calibration.

In this example, in order to use SPRT S/N 3447 for ITS-90 temperature measurements, its ITS-90 calibration constants a_s and b_s must be identified. Since the most recent calibration of this SPRT was done by NIST in November 1989 to the IPTS-68, this data will be used.[2] This is acceptable since the "hotness" of the fixed points at which the SPRT was calibrated have not changed; only their assigned temperatures have changed with ITS-90.

First, it is necessary to determine W_{68} for the SPRT at the triple point of water and the freezing points of tin and zinc. This can be done by interpolation from the calibration table provided by NIST. From the table the following values are obtained:

$$R(0.0) = 25.5432 \ ohms$$

$$W_{68}(TP \ H_2O) = W_{68}(0.01°C) \quad = 1.00003986$$

$$W_{68}(FP \ Sn) \ = W_{68}(231.9681) = 1.89256159$$

$$W_{68}(FP \ Zn) \ = W_{68}(419.58) \ = 2.56846594$$

where $TP \ H_2O$ is triple point of water and $FP \ Sn$ and $FP \ Zn$ are the freezing point of tin and zinc respectively.

Since the only difference between W for ITS-90 and IPTS-68 is whether the denominator is the triple point of water or the ice point resistance, conversion of W_{68} values to W_{90} are obtained by the following equation:

$$W_{90} = W_{68}/W_{68}(TP \ H_2O) \tag{9}$$

---

2. National Institute of Standards and Technology, "Report of Calibration-Platinum Resistance Thermometer-Rosemount Model 162CE-Serial No. 3447", November 9, 1989.

Therefore:

$$W_{90}(FP\ Sn) = W_{68}(FP\ Sn)/W_{68}(TP\ H_2O)$$

$$= 1.89256159/1.00003986$$

$$= 1.89248616$$

and

$$W_{90}(FP\ Zn) = W_{68}(FP\ Zn)/W_{68}(TP\ H_2O)$$

$$= 2.56846594/1.00003986$$

$$= 2.56836357$$

These values now correspond to the resistance ratios of the SPRT had they been measured relative to the SPRT resistance at the triple point of water rather than to the ice point. One additional change must be made to this data to make it appropriate for the ITS-90. The "hotness" of the tin and zinc points did not change; consequently the resistance of the SPRT will be the same at those points. The defined temperatures for these fixed points did change with the implementation of ITS-90. Using the ITS-90 temperatures for the fixed points their corresponding resistance ratios are as follows:

$$W(231.928°C) = 1.89248616$$

$$W(419.527°C) = 2.56836357$$

Using Equation 6, values of $W_r(T_{90})$ for the above two temperatures are obtained. Note that the $W_r(T_{90})$ values do not require any information about the SPRT calibration. The reference function values for the tin and zinc points are:

$$W_r(231.928°C) = 1.89279759$$

$$W_r(419.527°C) = 2.56891721$$

Since the values for the deviation function (Equation 8) at two temperatures have been obtained for the two constants, a_8 and b_8 can be calculated and are listed below:

$$a_8 = -3.43621777 \times 10^{-4}$$

$$b_8 = -5.89089256 \times 10^{-6}$$

These constants, a_8 and b_8, together with the resistance of the SPRT at the triple point of water provide a complete calibration of the SPRT for the ITS-90 subrange of 0 to 419.527°C.

To use this information to convert measured resistances of the SPRT to ITS-90 temperatures, one proceeds as follows:

1. Measure the resistance of the SPRT, $R(T_{90})$ at the unknown ITS-90 temperature.

2. Calculate the measured resistance ratio by:

$$W(T_{90}) = \frac{R(T_{90})}{R(TP\ H_2O)}$$

3. Calculate the value of the deviation function from:

$$\Delta W_8(T_{90}) = a_8[W(T_{90})-1] + b_8[W(T_{90})-1]^2$$

4. Calculate the value of the reference function at the unknown temperature from:

$$W_r(T_{90}) = W(T_{90}) - \Delta W(T_{90})$$

5. The unknown ITS-90 temperature is then calculated by using the specified inverse of Equation 7 which is:

$$T_{90} = D_o + \sum_{i=1}^{9} D_i \left(\frac{W_r(T_{90}) - 2.64}{1.64} \right)^i$$

Where the constant (D_o) and the coefficients (D_i) are tabulated in Reference 1.

In order to check the calculations used to convert the calibration of the SPRT S/N 3447 to the ITS-90, an "experiment on paper" was performed. Temperature measurements at a number of temperatures between 0 and 419°C (the melting point of Zinc) were assumed. These temperatures (t_{68}) and the IPTS-68 calibration of SPRT S/N 3447 were used to convert each of the (t_{68}) temperatures to the equivalent resistance which the SPRT would have at that temperature.

The ITS-90 calibration data calculated for SPRT S/N 3447 and the appropriate equations from ITS-90 were then used to convert the resistances to their equivalent ITS-90 temperatures. The difference between the IPTS-68 and ITS-90 temperatures were calculated at each "measurement" and compared to the differences listed in Reference 1. Figure 4 shows the differences between IPTS-68 and ITS-90 for SPRT S/N 3447. The points on the plot are the differences from Reference 1.

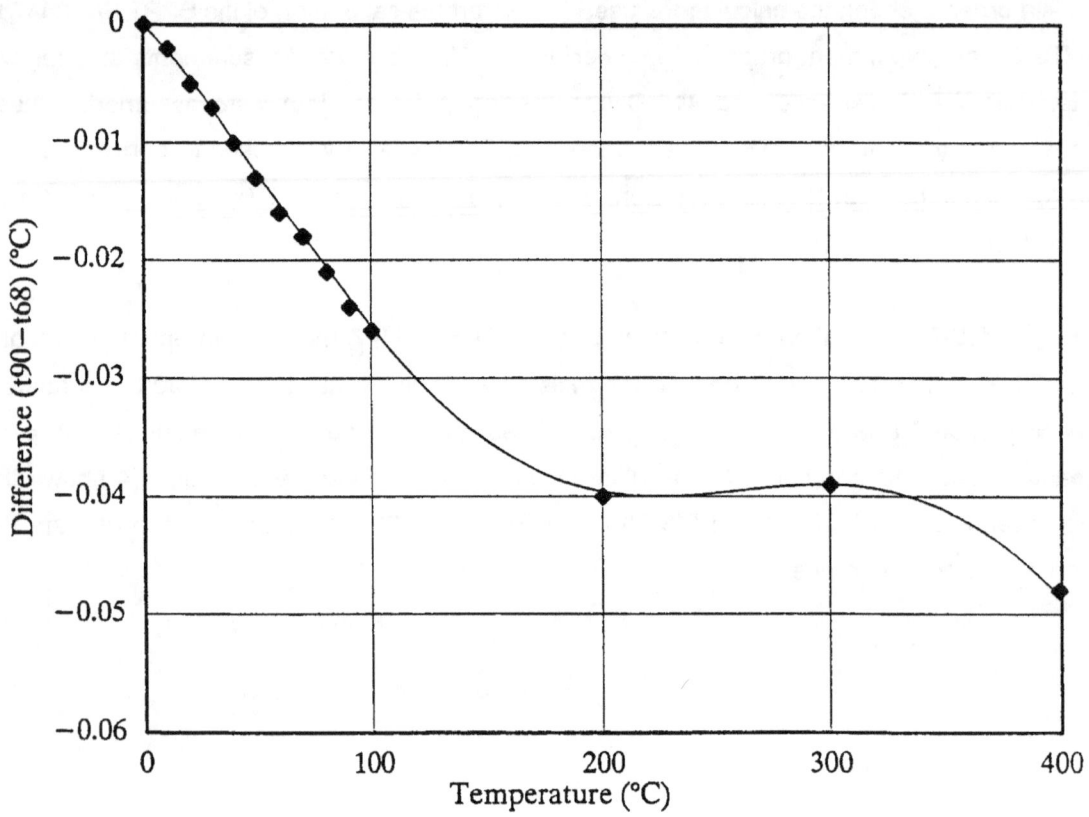

Figure 4.   Difference Between ITS-90 and IPTS-68
for SPRT S/N 3447.

NRC FORM 335
(2-89)
NRCM 1102,
3201, 3202

U.S. NUCLEAR REGULATORY COMMISSION

BIBLIOGRAPHIC DATA SHEET

(See instructions on the reverse)

1. REPORT NUMBER
(Assigned by NRC, Add Vol., Supp., Rev.,
and Addendum Numbers, if any.)

NUREG/CR-5560

2. TITLE AND SUBTITLE

Aging of Nuclear Plant Resistance Temperature Detectors

3. DATE REPORT PUBLISHED

MONTH	YEAR
June	1990

4. FIN OR GRANT NUMBER
D 2039

5. AUTHOR(S)

H. M. Hashemian, D. D. Beverly, D. W. Mitchell, K. M. Petersen

6. TYPE OF REPORT

Technical

7. PERIOD COVERED (Inclusive Dates)
October 1987 -
March 1990

8. PERFORMING ORGANIZATION — NAME AND ADDRESS (If NRC, provide Division, Office or Region, U.S. Nuclear Regulatory Commission, and mailing address; if contractor, provide name and mailing address.)

Analysis and Measurement Services Corporation
AMS 9111 Cross Park Drive
Knoxville, Tennessee 37923-4599

9. SPONSORING ORGANIZATION — NAME AND ADDRESS (If NRC, type "Same as above"; if contractor, provide NRC Division, Office or Region, U.S. Nuclear Regulatory Commission, and mailing address.)

Division of Engineering
Office of Nuclear Regulatory Research
U. S. Nuclear Regulatory Commission
Washington, D.C. 20555

10. SUPPLEMENTARY NOTES
Phase II Report

11. ABSTRACT (200 words or less)

An experimental research project was completed to identify the effects of normal aging on performance of nuclear safety-related RTDs. The limit for initial accuracy of these RTDs was established and the range of their response time was determined. Representative nuclear grade RTDs were tested for calibration drift at simulated reactor conditions and for shelf-life drift. This included a number of naturally aged RTDs received from nuclear power plants.

The results of this work have shown that periodic calibration and response time testing performed once every fuel cycle is a reasonable approach for management of aging of nuclear grade RTDs.

12. KEY WORDS/DESCRIPTORS (List words or phrases that will assist researchers in locating the report.)

Resistance Temperature Detectors (RTDs)
Aging
Calibration
Response Time
Drift
Accuracy
Degradation
Safety-related
Nuclear Plant

13. AVAILABILITY STATEMENT
Unlimited

14. SECURITY CLASSIFICATION

(This Page)
Unclassified

(This Report)
Unclassified

15. NUMBER OF PAGES

16. PRICE

NRC FORM 335 (2-89)